$10^8/k$ — количество организмов на гекта

ионарного числа (§ 37), а k — коэффициент

$$A_1 = \frac{pv^2}{2} \cdot \frac{10^8}{k} = \frac{pv^2 \cdot N_{\max}}{2 \cdot 5{,}10065 \cdot 10^{18}}.$$

резвычайно характерно, что эта величина
постоянная. Для них всех выражение

$$A_1 = \frac{pv^2}{2} \cdot \frac{10^8}{k} = a \cdot 3{,}51 \cdot 10^{12} \text{ C.G.S.,}$$

a — коэффициент, близкий к единице[1].

Iз этой формулы ясно, что кинетическая
:ozoa определяется скоростью v, связанной
анизма и с темпом размножения Δ. Отнес
дующей простой формулой [22]:

$$= \frac{46383{,}93 \cdot \lg 2 \cdot \Delta}{18{,}70762 - \lg k},$$

оторой числовые, постоянные для всех в
иенты связаны с размерами планеты (с
н с длиной ее экватора, причем все явлен
и секундам)[2].

не можем их понять, если н
планетных явлений, не обратим
рии, к ее атомам, к их измене
области быстро накапливают
ченные теоретической мыслью
аваться. Они не всегда могут б
ны, и выводы из них обычно не
о огромное значение этих явле
е факты должны теперь же учи
ях. Три области явлений мог
собое положение элементов зе
е, 2) их сложность и 3) неравном
ак, в массе земной коры рез
ы, отвечающие четным атом
снить это явление геологическ
не можем. К тому же нем
е явление выражено еще более
е космических тел, доступны
ению — для метеоритов (Гарки
бласть других фактов являет

e^2 я пытался выяснить. что

о крайней мере в 10 раз

ше 0,25% солнечной энер

в запасе — в живом ве

особом термодинамическ

поля косной материи био

оличества вещества, пост

ганизмы, большие колич

кислорода (около $1,5 \times 10$

ыражается в меньших чи

навливающиеся размнож

м, как указывалось (§ 45

ов, во много раз превыша

о, многократные числа

ас его учесть. энергетичес

организмо

$7)$, а k — коэ

$$pv^2 \cdot N_{max}$$

$\cdot 5{,}10065 \cdot 10$

оно, что эта

THE BIOSPHERE

THE BIOSPHERE VLADIMIR I. VERNADSKY

Foreword by
Lynn Margulis
Mauro Ceruti
Stjepko Golubic
Ricardo Guerrero
Nubuo Ikeda
Natsuki Ikezawa
Wolfgang E. Krumbein
Andrei Lapo
Antonio Lazcano
David Suzuki
Crispin Tickell
Malcolm Walter
Peter Westbroek

Introduction by
Jacques Grinevald

Translated by
David B. Langmuir

Revised and Annotated by
Mark A. S. McMenamin

A Peter N. Nevraumont Book

COPERNICUS
An Imprint of Springer-Verlag

Library of Congress Cataloging-in-Publication Data
Vernadskiĭ, V. I. (Vladimir Ivanovich), 1863–1945.
    [Biosfera. English]
    The biosphere/by Vladimir I. Vernadsky; forward by Lynn
 Margulis and colleagues; introduction by Jacques Grinevald;
 translated by David B. Langmuir; revised and annotated by Mark A.S.
 McMenamin.
       p. cm.
    "A Peter N. Nevraumont book."
    Includes bibliographical references.
    ISBN 0-387-98268-X (alk. paper)
    1. Biosphere.   I. McMenamin, Mark A.   II. Title.
 QH343.4.V4713   1997
 333.95—dc21   97–23855

Published in the United States by Copernicus,
an imprint of
Springer-Verlag New York, Inc.

Copernicus
Springer-Verlag New York, Inc.
175 Fifth Avenue
New York, New York 10010

A Peter N. Nevraumont Book

Manufactured in the United States of America
Printed on acid-free paper.
Designed by José Conde, Studio Pepín, Tokyo

Photograph of V.I. Vernadsky on jacket and
pages 4–6 courtesy of Mark McMenamin
0 9 8 7 6 5 4 3 2 1

ISBN 0-387-98268-X   SPIN 10557091

Produced by Nevraumont Publishing Company
New York, New York

President: Ann J. Perrini

CONTENTS

Foreword to the English-Language Edition

Just as all educated westerners have heard of Albert Einstein, Gregor Mendel, and Charles Darwin, so all educated Russians know of Vladimir Ivanovich Vernadsky (1863-1945). He is widely celebrated in Russia and the Ukraine. A Vernadsky Avenue in Moscow is rivaled by a monument in his memory in Kiev. His portrait appears on Russian national stamps, air letters, and even memorial coins.[1] A mineralogist and biogeologist, Vernadsky maintained his scholarly activity and laboratory at Moscow University through much revolutionary turmoil. He resigned, with most other scientists, in 1911. His previous part-time employment as adjunct member of the Academy of Sciences and director of the Mineralogical Museum in St. Petersburg permitted him to take on these activities full-time by the end of that year.[2]

When World War I broke out Vernadsky was leading an expedition to seek radioactive minerals in Siberia; by May 1917 he was elected head of his old department of Mineralogy and Geology at Moscow University. Ill health forced him to travel south and settle with his family in the Ukraine, until he left for Paris. He returned to the Soviet Union from France in 1926 and remained there until his death in 1945. Through the political morass of the Stalinist Soviet Union, Vernadsky remained vigilant towards honesty, indifferent to politics, and devoted to open scientific inquiry. The last part of his long career was also immensely productive: he continued to publish, lecture, attend conferences, organize institutes, express his opinions—popular or not. Indeed Vernadsky's entire life was dedicated to fostering the international scientific enterprise. His lectures at the Sorbonne in 1922-23 were known to Pierre Teilhard de Chardin and Edouard Le Roy. They were published in 1924 under the title *La Géochimie*.[3] In 1926 his greatest work, *The Biosphere*, the first full English translation of which you now hold in your hands, was published in Russian in 1926.[4] Vernadsky with great help from his wife and French colleagues then prepared a French language edition in 1929.[5] As only the most important and seminal

[1] In reviewing Bailes' book, Stephen M. Rowland of the University of Nevada, Las Vegas, writes "…Vernadsky symbolizes personal integrity and Slavic native ability. In the years to come, as the Russian and Ukrainian people look for sources of cultural pride, Vernadsky's stature is certain to grow. Already named in his honor are a mineral (vernadite), a geologic museum, the Ukrainian central science library, several mountain peaks and ranges, a peninsula in East Antarctica, a submarine volcano, a crater on the back side of the moon, a mine in Siberia, a scientific research vessel, a steamship, a village in Ukraine (Vernadovka), a street in Moscow (Vernadsky Prospekt), and a species of diatoms." See Rowland, 1993.

[2] Bailes, 1990.

[3] Vernadsky, 1924.

[4] Vernadsky, 1926a.

[5] Vernadsky, 1929.

scientific works are able to do, *The Biosphere* has remained fresh and current for over a half-century.

6 Seuss, 1875.

The descent of what Winston Churchill called the "iron curtain" and the subsequent cold war substantially reduced the flow of scientific information from Russia to Western Europe and beyond. This prolonged trans-European barrier deprived English-speaking scientific audiences of Vernadsky's imaginative and insightful books for most of this century. Vernadsky's obscurity in the West is surely one of the great examples in history of a political impediment to the spread of scientific information. But like the periodic table of elements, which, in the United States, is still seldom credited to its Russian inventor, Dimitri Mendeleev, Vernadsky's ideas became widely known even though they were not attributed to their author.

Although the Viennese geologist Eduard Seuss (1831-1914) had coined the term *biosphere* for the place on Earth's surface where life dwells,[6] and the word has since been used in various contexts by many scientists, it is Vernadsky's concept of the biosphere, as set forth in this book, that is accepted today. Three empirical generalizations exemplify his concept of the biosphere:

1 Life occurs on a spherical planet. Vernadsky is the first person in history to come grips with the real implications of the fact that Earth is a self-contained sphere.
2 Life makes geology. Life is not merely *a* geological force, it is *the* geological force. Virtually all geological features at Earth's surface are bio-influenced, and are thus part of Vernadsky's biosphere.
3 The planetary influence of living matter becomes more extensive with time. The number and rate of chemical elements transformed and the spectrum of chemical reactions engendered by living matter are increasing, so that more parts of Earth are incorporated into the biosphere.

What Vernadsky set out to describe was a physics of living matter. Life, as he viewed it, was a cosmic phenomenon which was to be understood by the same universal laws that applied to such constants as gravity and the speed of light. Still, Vernadsky himself and many of his fundamental concepts remained largely unknown.

His ideas first began to enter postwar Western science in the form of hybrid activities called "biogeochemistry," "geomicrobiology," or studies of ecosystems, ecology, and "environmental chemical cycles." Observation and measurement today of the flow of carbon, sulfur, and nitrogen through the hydrosphere, lithosphere, atmosphere, and biota are practices based on the style of thought invented by Vernadsky.

A statement of the major themes of V.I. Vernadsky's life work was coaxed from him by the great English-American ecologist G. Evelyn Hutchinson (1903-1991), of Yale University. Translated with the aid of Vernadsky's son, George Vernadsky, who taught history and Slavic studies also at Yale, it was organized into two articles, which are among the last published before Vernadsky's death in 1945, and they remained for many years the only pieces of his writing readily available to English-speaking readers.[7] Hutchinson's own chapter in *The Earth as a Planet*, one volume in the Dutch-American astronomer G. P. Kuiper's work on the solar system, itself embodied a Vernadsky-style conceptual shift. Our Earth in this scholarly encyclopedia is described as but one of nine planets, with life as its sole source of geochemical uniqueness.[8] Even James E. Lovelock, FRS, the British inventor and the other major scientific contributor to the concept of an integrated biosphere in this century, remained unaware of Vernadsky's work until well after Lovelock framed his own Gaia hypothesis.[9] Whereas Vernadsky's work emphasized life as a geological force, Lovelock has shown that Earth has a physiology: the temperature, alkalinity, acidity, and reactive gases are modulated by life. With the completion in 1996 of more than fifty SCOPE volumes on biogeochemistry and UNESCO's "Man and the Biosphere" program,[10] the word "biosphere" has clearly entered common parlance.

Demand for the voice of Vernadsky himself in English was given a boost by the Biosphere 2 project. The goal of this venture, financed by the Texas oil millionaire Edward Bass and run by a small, intensely private group of entrepreneurs, the "Ecotechnicians," was a completely self-sustaining living system within a 3.15-acre "greenhouse-with-an-ocean" in the Arizona desert just north of Tucson. In 1990, "Biospherians" in red space suits locked themselves into their gas-tight greenhouse for a planned two-year stay; they encountered diminishing oxygen supplies, dangerously high concentrations of carbon dioxide, disastrous "extinctions" of many species, and even more disastrous population explosions of others. By 1992 the structure was opened, the "ecologically closed life support-system experiment" ended, and the facility at Oracle, Arizona, was

7 Vernadsky, 1944, 1945.

8 Hutchinson, 1954.

9 Lovelock, 1988.

10 Munn, 1971-1996.

made available to others for scientific research. This facility, presently the largest greenhouse in the world, is currently administered by Columbia University of New York City. Wallace S. Broecker, Director of the Lamont Geophysical Laboratory in Palisades, New York, which houses the geology department of Columbia, describes in a lively account the history and current status of Biosphere 2 of which he was the first director.[11] The current president and executive director of Biosphere 2 is the former assistant director of mathematics and physical science at the US National Science Foundation; William Harris.

In search of both financial support and philosophical guidance, Biosphere 2's publishing arm, Synergetic Press, in 1986 published an 83-page bowdlerized translation of Vernadsky's *The Biosphere*, based on the 232-page French text.[12] When one of us (LM) asked editor-in-chief Tango Snyder Parrish (a.k.a Deborah Snyder) why the book was so thin, she replied that she had removed everything that might, in retrospect, be considered "wrong" and so might blemish Vernadsky's posthumous reputation. This unconscionable mangling further frustrated the tiny readership that now clamored for the real Vernadsky.

Biospherians notwithstanding, the rediscovery of Vernadsky was by this time underway. An excellent account of the scientific insights of Vernadsky, his colleagues and students, *Traces of Bygone Biospheres*[13] was published by Leningrad geochemist Andrei Lapo. This small book became available in English in 1987[14] and later also was distributed by Synergetic Press.[15] Kendall Bailes' magnum opus,[16] the definitive English-language biography of Vernadsky and his times, appeared posthumously in 1990. That Vernadsky had written widely, that his name had been honored by several scientific institutions, awards, and publications in the USSR, and that he was the first to recognize the importance of life as a geological force,[17] were more and more widely discussed in a multitude of languages.[18] It was primarily Jacques Grinevald's presentation at the first Gaia meeting in 1987, organized by Edward Goldsmith and Peter Bunyard and published in a devilishly difficult to obtain book,[19] that made us all painfully aware of the main problem: the authentic Vernadsky remained unavailable to an English-language readership. A newly published version of the Gaia meeting and its successors is now available.[20]

For at least a decade prior to the appearance of the Synergetic Press pamphlet, a 187-page typescript of the entire *The Biosphere* in English translation was circulating in Boston and New

11 Broecker, 1996.

12 Vernadsky, 1986.

13 Lapo, 1979.

14 Lapo, 1982.

15 Lapo, 1987. A third edition is planned to be published by Copernicus/Springer Verlag in 1999.

16 Bailes, 1990.

17 Westbroek, 1991.

18 Tort, 1996.

19 Grinevald, 1988.

20 Bunyard, 1996.

York. Its mysterious, nearly blank title page offered no information about its origins other than that it had been "translated by David Langmuir." Lynn Margulis received a copy from her colleague, Thomas Glick, of the History and Geography Departments at Boston University, who had taught courses on the impact of Charles Darwin.[21] She enthusiastically read and passed on the by now well-worn typescript to her former student Betsey Dexter Dyer, who currently teaches biology at Wheaton College. Dyer had aided Andrei Lapo in preparing the English translation of the second edition of his *Traces* while she enjoyed a U.S. National Academy of Sciences-sponsored research excursion to the Soviet Union in 1984 and now she requested his help in locating the translator. In the end both Lapo and Jacques Grinevald provided the same address. Our enterprising publisher, Peter N. Nevraumont, found Langmuir in Santa Monica, California, alive and well in his mid-eighties, with his faculties fully intact. Langmuir, of course, was delighted to learn that a full English translation of Vernadsky's *The Biosphere* would at long last see the light of day.

Even without an accessible version of his greatest book, we have all felt Vernadsky's influence on our work. Indirectly through Lapo, Bailes, Grinevald, the red-suited biospherians, the two articles sponsored by Hutchinson in the 1940s, the writings of M. M. Kamshilov,[22] Westbroek, the legacy of G. E. Hutchinson,[23] and for those of us who could locate it, A. E. Fersman's wonderful book *Geoquímica Recréativa*, released in English as *Geochemistry for Everyone*,[24] we have been informed in many ways of Vernadsky's ideas. Our debt now to Peter N. Nevraumont for his willingness to spread the word is immeasurable. A world-class scientist and writer, Vernadsky needs no protection from the guardians of the politically correct, whether biospherians, anti-Communists, or others. Vernadsky is finally allowed to speak in English for himself.

Vernadsky teaches us that life, including human life, using visible light energy from our star the Sun, has transformed our planet over the eons. He illuminates the difference between an inanimate, mineralogical view of Earth's history, and an endlessly dynamic picture of Earth as the domain and product of life, to a degree not yet well understood. No prospect of life's cessation looms on any horizon. What Charles Darwin did for all life through time,[25] Vernadsky did for all life through space. Just as we are all connected in time through evolution to common ancestors, so we are all—through the atmosphere, lithosphere,

21 Glick, 1974.

22 Kamshilov, 1976.

23 Hutchinson, 1957-1992.

24 Fersman, 1958.

25 Darwin, 1963 [first published 1859].

hydrosphere, and these days even the ionosphere—connected in space. We are tied through Vernadskian space to Darwinian time.[26] We embrace the opportunity afforded by Copernicus Books of Springer-Verlag to, at long last, cast broadly the authentic Vernadskian English-language explanations of these connections.

26 Margulis, L. and D. Sagan. 1995.

Lynn Margulis   University of Massachusetts, Amherst, Ma., USA

Mauro Ceruti   Department of Linguistics and Comparative Literature, University of Bergamo, Italy

Stjepko Golubic   Boston University, Boston, Ma., USA, and Zagreb, Croatia

Ricardo Guerrero   Department of Microbiology, University of Barcelona, Spain

Nubuo Ikeda   Graduate School of Media and Governance, Keio University, Japan

Natsuki Ikezawa   Author (*Winds From the Future*; *The Breast of Mother Nature*—Yomiuri Bungaku Award) Japan

Wolfgang E. Krumbein   Department of Geomicrobiology, University of Oldenburg, Germany

Andrei Lapo   VSEGEI (All-Russian Geological Research Institute), St. Petersburg, Russia

Antonio Lazcano   Department of Biology, Universidad Autonoma Nacional de Mexico, Mexico

David Suzuki   University of British Columbia and Canadian Broadcasting Company, Canada

Crispin Tickell   Green College, Oxford, United Kingdom

Malcolm Walter   School of Earth Sciences, Macquarie University, Sydney, Australia

Peter Westbroek   Department of Biochemistry, University of Leiden, The Netherlands

Introduction: The Invisibility of the Vernadskian Revolution

I suggest that there are excellent reasons why revolutions have proved to be so nearly invisible. Both scientists and laymen take much of their image of creative scientific activity from an authoritative source that systematically disguises—partly for important functional reasons—the existence and significance of scientific revolutions.

Thomas S. Kuhn
The Structure of Scientific Revolutions[27]

27 Kuhn, 1962, p. 136.

28 Hutchinson, 1965, pp. 1-26; and 1970.

In his epoch-making article introducing the September 1970 issue of *Scientific American* devoted to "the Biosphere," the founder of the Yale scientific school in ecology, George Evelyn Hutchinson, wrote :

> The idea of the biosphere was introduced into science rather casually almost a century ago by the Austrian geologist Eduard Suess, who first used the term in a discussion of the various envelopes of the earth in the last and most general chapter of a short book on the genesis of the Alps published in 1875. The concept played little part in scientific thought, however, until the publication, first in Russian in 1926 and later in French in 1929 (under the title *La Biosphère*), of two lectures by the Russian mineralogist Vladimir Vernadsky. It is essentially Vernadsky's concept of the biosphere, developed about 50 years after Suess wrote, that we accept today.

Hutchinson's authoritative assessment[28] has not been fully appreciated. For most people in the West, the name Vladimir Ivanovich Vernadsky (1863-1945) is still largely unknown. In fact, apart from rare exceptions like the entry in the multi-volume *Dictionary of Scientific Biography*, most of our usual reference books, including those in the history of science, ignore Vernadsky and his Biosphere concept. The world's first scientific monograph on the Biosphere of Earth as a planet, which Vernadsky published in 1926, is not yet listed among the major books that have shaped our modern world view.

The special issue of *Scientific American* on "the Biosphere," a landmark in all respects, was published at the beginning of "the environmental revolution," to borrow the title of Max Nicholson's 1970 book. In Western industrial societies, this epoch was marked by the political emergence of a global environmental movement, internationally recognized at the 1972 United Nations Stockholm Conference. Following the so-called "Biosphere Conference"[29] organized by UNESCO, Paris, in September 1968, the world *problematique* of "Man and the Biosphere" (UNESCO's MAB program) became a pressing issue for many of us, reviving, either explicitly or implicitly, views on the biosphere that had originated with Vernadsky.[30]

The Vernadskian renaissance began slowly, in the 1960s and 1970s in the Soviet Union, thanks to a little circle of scholars within the Academy of Sciences. By the time of Gorbachev's perestroika, Vernadsky was a cult figure for the liberals and a national icon for others. With the collapse of the USSR a major barrier to the official recognition of Vernadsky's life and work came down as well.[31] The international revival of Vernadsky came of age in the mid-1980s; many circumstances, including the Biosphere 2 project, are recalled in the forward to this volume.

Paradoxically, at the birth of the environmental era, the very term biosphere was often betrayed or replaced, for example, the vague notion of "global environment." In another instance, the biosphere was correctly named in *Science and Survival* (1966) by Barry Commoner, but in his international best seller *The Closing Circle* (1971), *biosphere* was unfortunately replaced by *ecosphere*. This neologism, in vogue since 1970, was introduced in flagrant ignorance of Vernadsky's teaching. With the use of the *ecosphere*, the concept of biosphere was reduced to the "global film of organisms."[32] This is a far narrower, more pedestrian idea than what Vernadsky proposed.

In the 1970s and 1980s, many Soviet publications on global environmental issues, including nuclear war, praised Vernadsky as the originator of the modern theory of the Biosphere. One of the first works to do so, the textbook *Global Ecology*, written by the Soviet climatologist Mikhail I. Budyko, was published in English and French by 1980.[33] Involved in the global warming debate since the early 1970s, Budyko was an internationally known meteorologist and the author of several books on climatic aspects of "the evolution of the Biosphere."[34] But as

29 *Use and Conservation of the Biosphere*, 1970.

30 Vernadsky, 1945.

31 Tort, 1996, pp. 4439-4453.

32 Lieth and Whittaker, 1975.

33 Budyko, 1980.

Kendall Bailes, the author of the only English-language biography of Vernadsky, pointed out, Soviet appraisals often merged with the official ideology, so that the life and thought of Vernadsky were often horribly distorted.[35] French and English versions of the beautiful, though not always reliable, monograph *Vladimir Vernadsky*, by Rudolf Balandin, appeared in the series "Outstanding Soviet Scientists."[36] Semyon R. Mikulinsky, with the Institute of the History of Science and Technology of the Academy of Sciences, emphasized the neglected work of Vernadsky as an historian of science, but still with an obvious communist slant.[37]

In the early 1980s, Nicholas Polunin, writing in the international journal *Environmental Conservation*, emphasized "the wide-prevailing public ignorance concerning the Biosphere, which is such that the vast majority of people living in it (as of course all people normally do) simply do not understand what it is, much less realize how utterly dependent they are on it for their life-support and very existence."[38] Polunin was a British plant geographer turned environmentalist and a former collaborator at Oxford with Arthur Tansley, the British botanist who in 1935 coined the term "ecosystem." It was Polunin who proposed the convention of writing "Biosphere," in the sense of Vernadsky, with a capital letter, to emphasize the unique standing of the only living planet we know in the cosmos. It is also useful to distinguish it from the other meanings, including the biosphere as a part of the climate system.[39] Polunin defined the Biosphere as the "integrated living and life-supporting system comprising the peripheral envelope of planet Earth together with its surrounding atmosphere so far down, and up, as any form of life exists naturally."

The invisibility of the Vernadskian revolution is part of the cultural history of ideas. From the start, the term *Biosphere* has been interpreted in many different and contradictory ways. A scientific consensus on the term is still lacking.[40] The scientific concepts of Vernadsky compete with and are frequently superseded by other popular terms and ideas, including Teilhardism, the worldwide cultural movement accompanying the posthumous edition of Teilhard de Chardin's writings on science and religion. Teilhard developed his own notions of "Biosphere" in many fascinating texts, but not in his strictly scientific works,[41] though it was not a clear-cut division for him. Even noted authors erroneously credited Teilhard de Chardin for the word

34 Budyko, 1986.

35 Bailes, 1990; and Yanshin and Yanshina, 1988.

36 Balandin, 1982.

37 Mikulinsky, 1984; and 1983, with a commentary by Tagliamgambe.

38 Polunin, 1972; 1980; 1984; Polunin and Grinevald, 1988.

39 As stated in the definitions (Article 1) of the United Nations Framework Convention on Climate Change (1992), as well as in many official scientific publications, including the authoratative reports of Intergovernmental Panel on Climate Change (IPCC).

40 See the summary of state of art by the British geographer Richard J. Huggett, 1991; and 1995.

41 Teilhard, 1955-1976; and 1971.

"biosphere,"[42] though both Teilhard and Vernadsky were careful to attribute it to the great Austrian geologist Eduard Suess (1831-1914).[43] It is equally misleading, of course, to state that Vernadsky originated the term. This mistake, frequently made since UNESCO's 1968 conference, even appears in Peter J. Bowler's *History of Environmental Sciences*.[44] It is a flagrant illustration of the widespread ignorance of Vernadsky's own writings, as well as of the history of the idea of the Biosphere.[45] Sometimes, Teilhard and Vernadsky are merged, as, for instance, in Theodosius Dobzhansky's 1967 book *The Biology of Ultimate Concern*.[46] As Thomas F. Malone wrote:

> The proposal to unite geophysics and biology is the culmination of conceptual thinking that began in 1875 with the identification of the "biosphere"—described by the Suess—as the concentric, life-supporting layer of the primordial Earth. It has been developed as a concept in modern scientific thought largely through the work of Vernadsky during the 1920s.[47]

The Suessian model of geological envelopes,[48] or "geospheres" (the term coined by *Challenger* oceanographer John Murray in 1910), was adopted by geographers, meteorologists (*troposphere* and *stratosphere* were introduced by Léon Teisserenc de Bort in 1902), geophysicists (*asthenosphere* was introduced by Joseph Barrell in 1914), and soil scientists (*pedosphere* was coined by Svante E. Mattson in 1938). This scheme of geospheres gained wide currency through the three great founding fathers of modern geochemistry, the American Frank W. Clarke (1847-1931), chief chemist to the U.S. Geological Survey (1883-1925);[49] the Zurich-born Norwegian geologist Victor Moritz Goldschmidt (1888-1947), whose the life was disturbed by Hitler's accession to power;[50] and Vernadsky. As the founder of the Russian school of geochemistry and biogeochemistry, Vernadsky was mentioned in the major books on geochemistry when that field came of age after World War II.[51] Then, apparently, the name of Vernadsky was forgotten, as philosophers and historians of science neglected the growing role of Earth and planetary sciences in contemporary scientific knowledge. The case of Vernadsky is, of course, not unique in the history of Soviet science.[52]

Between Suess and Vernadsky, a pioneering movement helped to merge biology and geology but, as always, the beginnings were obscure. Led by the German naturalist Johannes

42 For instance, the great historian Arnold Toynbee (1889-1975). See Toynbee, 1976 (chapter 2: "The Biosphere").

43 Vernadsky, 1929 (§68). Reference to E. Suess, 1875, p. 159: "One thing seems strange on this celestial body consisting body consisting of spheres, namely organic life. But this latter is limited to a determined zone, at the surface of the lithosphere. The plant, which deeply roots plunge in the soil to feed, and at the same time rises into the air to breathe, is a good illustration of the situation of organic life in the region of interaction between the upper sphere and the lithosphere, and on the surface of continents we can distinguished a self-maintained biosphere [eine selbständige Biosphäre]."

44 Bowler, 1992.

45 Grinevald, 1988.

46 On Dobzhansky's connection with both Vernadsky and Teilhard, see Adams, 1994.

47 Malone and Roederer, 1985, p. xiii.

48 Vernadsky, 1924, pp. 64-74, "Les enveloppes de l'écorce terrestre."

49 Since 1909, Vernadsky read the successive editions of Clarke's *Data of Geochemistry*. In the second edition (1911), Vernadsky is quoted ten times.

50 Mason, 1992, contains Goldschmidt's complete bibliography. Plate 23 of this authoratative book is a photograph of Goldschmidt and Vernadsky in front of Goldschmidt's home in Göttingen in June 1932.

51 Rankama and Sahama, 1950; and Mason, 1952. These two books include some useful historical information, but the complete history of geochemistry (and its role in the rise of Earth sciences), since the coinage of its name in 1838 by the German chemist Christian Friedrich Schönbein (1799-1868), the discoverer of ozone, professor at the University of Basel (Switzerland), has yet to be written.

52 Graham, 1993.

Walther (1860-1937); Stanlislas Meunier (1843-1925), the author of *La Géologie biologique* (1918); and the Harvard physiologist Lawrence J. Henderson (1878-1942), the author of *The Fitness of the Environment* (1913). The sources of Vernadsky are in fact an immense library, which is the intellectual prehistory of Gaia. As Alexander P. Vinogradov wrote on the occasion of the 100th anniversary of Vernadsky's birth: "Much time will have to pass before the historian of science will be able to review the vast scientific legacy of Vernadsky and fully grasp the depth and many-sidedness of his influence."[53] The same is true for the historical sources of Vernadsky's work.

In the English-speaking world, the idea and term biosphere were not quickly diffused, or else were used in the restricted sense given by the geochemists.[54] In France, just after World War I and the publication (delayed because the war) of the final volume of *La Face de la Terre*, the little French Catholic circle of geologists and naturalists with the Muséum d'Histoire Naturelle enthusiastically adopted Suess's notion of "biosphere," while rejecting Wegener's revolutionary theory of continental drift. The geologist Pierre Termier spoke of Wegener's idea as "a dream, the dream of a great poet," a "seductive," "unnecessary," "extremely convenient" hypothesis (the same words used by Lovelock's opponents!).[55] Teilhard de Chardin, already a noted scientist as well as a mystic prophet, adopted Suess' biosphere when he finished reading *La Face de la Terre* in 1921.[56] When he met Vernadsky, he was completely ignorant of the biogeochemical approach developed by his Russian colleague. Both Vernadsky and Teilhard praised Suess for the term *biosphere* and saw the need of a new "science of the Biosphere."[57] But the common terminology is misleading. In fact, the French scientist and the Russian scientist were interpreting *biosphere* in radically different ways. It is interesting to note that Teilhard developed his own notion of "Biosphere" mainly in his philosophical writings,[58] not through his scientific publications.[59] For his part, Vernadsky based his evolving conception of the Biosphere essentially in his biogeochemical works, but also in his work in philosophy of science, including a concept he termed "empirical generalizations." Like many great scientists of his time, Vernadsky developed his personal philosophical thought on the great questions. Like Teilhard within his order of the Society of Jesus, our Russian scientist was censored and his intellectual activities restrained during the Stalinist era.

Both Vernadsky and Teilhard were cosmic prophets of global-

[53] Vinogradov, 1963, p.727.

[54] Goldschmidt, 1929; and Mason, 1954.

[55] Termier, 1915; 1922; 1928; 1929; and Termier and Termier, 1952.

[56] Teilhard, 1957a.

[57] Teilhard, 1957b.

[58] Teilhard, 1955-1976.

[59] Teilhard, 1971.

ization. If Teilhard was a "cosmic mystic," Vernadsky defined himself as a "cosmic realist." They shared a belief in science and technology as a universal, peaceful and civilizing force. Energy (force, power, work, production) was the key-word of the *Zeitgeist*. Vernadsky and Teilhard both offered energetic interpretations (but based on different energetics) of biological and technological systems shared then by the Ostwaldian, Machian and Bergsonian thinkers, extending the ideas of energetics and biological evolution to human "exosomatic evolution" (a term later coined by A. Lotka).[60] But in *The Biosphere* as in all his work, Vernadsky's scientific perspective is radically different from that of Teilhard. The divergence is perhaps best expressed as an opposition between the anthropocentric view of life (Teilhardian biosphere) and the biocentric view of the nature's economy (Vernadskian Biosphere).

Vernadsky's Biosphere is completely different from what Gregory Bateson called the "Lamarckian biosphere" of Teilhard, which was still connected to the classical idea of the Great Chain of Being and not at all synonymous with our modern ecosystem concept. I suspect Teilhard (long in China) did not read Vernadsky's *La Biosphère*. It was never cited in all Teilhard's published letters, philosophical writings and scientific works. Teilhard is not alone. Even after Vernadsky's death, *The Biosphere* was mentioned, if at all, in the necrological notices with a curious sort of no comment.

As an analytical abstraction for studying the complexity of nature, the functional concept of ecosystem, formally introduced after Vernadsky's *The Biosphere*, has no geographical boundary outside the observer's choice. Its extent is defined by the scale of observation. To quote Vernadsky, "there is nothing large or small in nature."[61] If Earth seems small to us now, it is because man's power, a manifestation of conscious life in the evolving Biosphere, is becoming large. While the biota, including microorganisms, constitutes a relatively small biomass compared with the total mass of the lithosphere, the hydrosphere and the atmosphere, the planetary role of living matter in nature's economy—to recall the classical metaphor at the roots of ecology—is enormous. According to Vernadsky, the Biosphere is not only "the face of Earth" but is the global dynamic system transforming our planet since the beginning of biogeological time. Vernadsky's position on the origin of life on Earth evolved, but I prefer to ignore Oparin's impact[62] on Vernadsky's opinion on genesis.[63] A comparative study is still to be written.

Vernadsky's long neglected discovery that the Biosphere, as

60 Lotka, 1945. A comparative study of Lotka and Vernadsky is still missing.

61 Vernadsky, 1930b, p. 701.

62 Oparin, 1957.

63 Vernadsky, 1932.

the domain of life on Earth, is a *biogeochemical* evolving system with a cosmic significance, was a scientific novelty unwelcomed by mainstream science. It was indebted to many new and old ideas in science, as well as in philosophy, Bergson's anti-mechanist epistemology of life notably. Vernadsky's Biosphere concept was part of the new geochemical point of view that considered Earth as a dynamic energy-matter organization, a system comparable to a thermodynamic engine. Following the insights of early bioenergetics, including the essay on metabolism (*Stoffwechsel*) published in 1845 by the German physician Robert Mayer,[64] the works of the German plant physiologist Wilhelm Pfeffer,[65] and the study on "the cosmic function of the green plant" by the Russian Darwinist Kliment A. Timiryazev,[66] Vernadsky viewed the Biosphere as, "a region of transformation of cosmic energy."[67] Energetics of the Biosphere, as Vernadsky emphasized, implies Earth systems as a planet functioning in the cosmic environment, powered by the Sun. This new thermodynamical cosmology was the result of what we have called the Carnotian revolution. At the beginning of the twentieth century, modern science was transformed by an explosion of discoveries and inventions. The microphysics of quanta and Einstein's theory of relativity were part of this profound metamorphosis. Biological sciences and earth sciences were also profoundly altered by developments in applied mathematics (the so-called probabilistic revolution), and the physical and chemical sciences. The engineering-born science of thermodynamics, connected with physiology, biochemistry, and (later) ecology, was pivotal in the emergence of the concept of Earth as an evolving system powered by internal and external energy sources. Earth system science was still embryonic during the age of Vernadsky, but the author of *The Biosphere* was thinking ahead of his time. One of Vernadsky's core ideas was "biogeochemical energy" (*The Biosphere*, §25). This energy-centered approach was clearly part of the second Scientific Revolution of the West, of which Bergson and Le Roy were early philosophers.

To a certain degree, the intellectual confusion surrounding the holistic idea of the Biosphere is the result of the mechanistic reductionist nature of Western mainstream science, as clearly expressed by the Cartesian philosophy of Jacques Monod's *Chance and Necessity.*[68] Other and opposite reasons exist, including Teilhard de Chardin's pervasive influence, as we have seen.[69] Mainstream science viewed holism as vitalist and antiscientific. Paradoxically, like Teilhard, one of his holistic villains,

[64] Vernadsky, 1924, pp.329-330, 334-338; and the English translation of Mayer, 1845, in *Natural Science*.

[65] Vernadsky, *The Biosphere* (§91); and Bünning, 1989.

[66] Timiryazev, 1903.

[67] Vernadsky, 1924, chap. III, §21, *The Biosphere* (§8); and Trincher, 1965 (containing a long extract of Vernadsky's *Geochemistry* on the Carnot principle and life, pp. 84-93). Compare these with Odum, 1971; Gates, 1962; and Morowitz, 1968.

[68] Monod, 1971.

[69] In the West, Teilhard de Chardin's extraordinary fame was practically inverse of that of Vernadsky. The debated figure of Teilhard de Chardin is still present, as illustrated by the recent acclaimed books (ignoring Vernadsky) of Barrow and Tipler, 1986; or Duve, 1995.

Monod, used the term "biosphere" only in the restricted sense of biota, ignoring Vernadsky's concept even though it had already been adopted by ecosystems ecology. Like Monod, many modern biologists and biochemists are ignorant of ecology and the Biosphere. Modern geochemistry, a "big science" in the nuclear age, is also guilty of neglecting the whole.[70] The four-box or reservoir scheme (atmosphere-hydrosphere-lithosphere-biosphere) still represents the dominant geochemical and geophysical paradigm.

The scientific awareness expressed by Vernadsky and some forgotten naturalists was long absent from mainstream science, and until recently was not a "global issue" of national politics or international affairs.[71] In it's new intellectual and international context, Vernadsky's scientific revolution is beginning to emerge from the haze of its early manifestations. The revolutionary character of the Vernadskian science of the Biosphere was long hidden by the reductionist, overspecialized and compartmentalized scientific knowledge of our time.

At the dawn of the twentieth century, after three centuries of "modern science," biology appeared in contradiction with the physical and chemical sciences. The paradox, pointed out by several authors, notably the French philosopher Henri Bergson in his great book *L'Evolution créatrice* (1907), was the famous second law of thermodynamics, the Carnot principle named the law of entropy. The anti-mechanistic philosophy of Bergson's *Creative Evolution* celebrated Life as an improbable diversity-creating whole, animated by a powerful *élan vital*, an accelerating biological coevolution transforming the inert matter of the surface of this planet. This had a profound influence on Vernadsky, as on many naturalists among the Russian intelligentsia before the Bolshevik Revolution. In the second part of his long scientific career (he was over 60 when he wrote *The Biosphere*), Vernadsky's intellectual ambition was to reconcile modern science with biological processes and life as a whole of cosmic significance.

The discovery of "the living organism of the biosphere" (*The Biosphere*, §13) arose from many scientific traditions and innovations integrated by the encyclopedic mind of Vernadsky, a geologist and naturalist in the broadest sense of the terms. Vernadsky made this intense intellectual effort, during a difficult but especially creative period from 1917 to the mid-1920s. It was an historical epoch marked by World War I, the Russian revolutions, and an extraordinary emotional context admirably

[70] For the leading geochemist V. M. Goldschmidt: "The totality of living organisms represents the biosphere, *sensu strictu*, and through its metabolism the biosphere is most intimitely connected with the atmosphere, hydrosphere, and pedosphere" (see Goldschmidt, 1954, p. 355). See also Mason and Moore, 1982, p.41: "The biosphere is the totality of organic matter distributed through the hydrosphere, the atmosphere, and on the surface of the crust." This definition goes back to the first edition of Mason's *Principles of Geochemistry*, 1952.

[71] See Grinevald, 1990; Smil, 1997.

described by Lewis Feuer's *Einstein and the Generation of Science*.[72] But, like all the scholarly literature on the scientific revolutions, Feuer's great book ignored Vernadsky and his revolutionary theory of the Biosphere.

Like the rest of science in the twentieth century, biology developed many subdisciplines and was deeply influenced by what E. J. Dijksterhuis called "the mechanization of the world picture."[73] The modern philosophy of the laboratory, in biology as well as in chemistry, produced a mechanistic and reductionistic science, more and more separated from evolving nature. Classical organized biology was often considered an outmoded science after the discovery of the structure of DNA in the 1950s. The biological and geological sciences developed as separate fields of knowledge, ignoring Vernadsky's integrative approach. In this modern institutional and epistemological context, Vernadsky's interdisciplinary and holistic concept of the Biosphere was a very unwelcome scientific idea.

After the initial success of the Newtonian-Laplacian world of mechanics and the atomistic reductionism of statistical mechanics, a holistic integration of life with the rest of the physical world was not fashionable. Some eminent scientists saw in holism the return to vitalism or even the older tradition of life *of* Earth, a sort of revival of the ancient mythology of Gaia! Thermodynamics was not a model for Suess's biosphere, but it was for Vernadsky and Lotka, much more than for Teilhard. Unfortunately, at the time, evolution was seen as opposed to the second law of thermodynamics.

Now we know that thermodynamics was itself an evolving science.[74] As early as the end of the nineteenth century, several experts thought that the entropy law was a law of evolution. Thermodynamics connects organisms with their environment, life with Earth, and Earth as a planet with its cosmic environment. The face of Earth, this strange "domain of life" in the cosmos, can be seen as an evolutionary phenomenon, the result of metabolism connecting the living organisms, the energy flow, and the cycling of chemical elements. Influenced by Bergson[75] this growing scientific awareness was first expressed by Vernadsky in parallel with Lotka, a well-recognized pioneer of the ecological worldview.

Like many forgotten energetist scientists, Vernadsky was seriously interested in the apparent thermodynamic paradox between life and entropy. The debate of Darwin versus Carnot was an epistemological problem with immense social implica-

72 Feuer, 1974.

73 Dijksterhuis, 1961

74 Wicken, 1987; Depew and Weber, 1995.

75 Vernadsky, 1934; and 1935, pp.208-213; 308-312.

tions. This debate was in fact a question of scale and boundary, viewpoint and measurement, in both space and time. At the time, it was not clearly recognized that the systems can be divided into three categories: isolated (no energy-matter exchange with the environment), closed (only energy exchange) or open (energy-matter exchange). In fact, the boundary of a phenomenon is an artifact defined by the observer. In emphasizing this point Vernadsky was criticizing not the productive division of scientific labor but the compartmentalization of scientific knowledge, and especially the loss of the unified view of nature shared by the great naturalists of the past.

When Vernadsky was writing on energetics of the Biosphere, energetics was a controversial epistemological matter with striking ideological implications, notably in Russia. Vernadsky's integrative and holistic framework, merging the anti-mechanistic approach of thermodynamics and the new atomism,[76] was a source for systems thinking in the Soviet Union.[77] The science of thermodynamics, linked with the engineering of the Industrial Revolution, physical chemistry as well as physiology and biochemistry, and later also ecology and environmental sciences, was considered at the end of the nineteenth century to present a new, anti-Newtonian scientific worldview. In a French paper of 1927, Vernadsky wrote:

> At the end of the last century, we witnessed the influence of the *energetic mentality* upon the scientific understanding of nature. This coincided with an effort to give a dynamical conception of environment, a conception that seemed in perfect harmony with the thermodynamic view of the Universe.[78]

This conception—generally credited only to Lotka in the United States—was the real beginning of the study of energy flows in ecosystems, "at various size levels ranging from simplified microcosms to the biosphere as a whole."[79] In fact, the epistemological conflict between Darwin and Carnot was not resolved until Schrödinger's *What is Life?*, which impelled a change in perspective and the emphasis on the point that what we call "structure" in biological organization is as much a self-organization of *processes* as of structures.[80] In *The Biosphere* (§89), Vernadsky emphasized that: "Our model of the cosmos always must have a thermodynamic component." It is in this context, prior to Schrödinger and Prigogine, that we can appreciate Vernadsky's warning expressed in the 1929 French edition of *The Biosphere*:

76 For a historical review of this epistemological conflict see Clark, 1976, pp.41-105.

77 Susiluoto, 1982, pp.25-27. There is an interesting parallel between Vernadsky's biospheric framework and the Russian thinker A. Bogdanov (pseudonym A. A. Malinovsky, 1873-1928), the author of the 3 volume treatise *Tektology The Universal Organizational Science* (1925-1929, 3rd ed., 1922, Moscow). See Bogdanov, 1980.

78 Vernadsky, 1927.

79 Odum, 1968. Vernadsky is also omitted by Gallucci, 1973. However, Vernadsky was quoted in Kuznetsov, 1965.

80 Bergson, 1975; Whitehead, 1926; and especially Georgescu-Roegen, 1971. See also Glansdorff and Prigogine, 1971; Denbigh, 1975; and Weber, 1988.

"The physical theories must inevitably be preoccupied with the fundamental phenomena of life" (*La Biosphère*, 1929, appendice, p.229).

To illustrate this new trend of scientific thought, Vernadsky cited J. Lotka's *Elements of Physical Biology* (1925), A. Whitehead's *Science and Modern World* (1926) and J.B.S. Haldane's *Daedalus* (1924). These three references form a useful framework for the critical study of *The Biosphere*. Parallel to Lotka, Vernadsky emphasized the human implications of energetics of our living Earth, its ecological limits and economic possibilities. Unfortunately, mainstream biologists and economists long ignored this bio-economic message until the revolutionary work of the unorthodox economist Nicholas Georgescu-Roegen, who showed the way for the critique of the mechanistic dogma of neoclassical economics. Both Vernadsky and Georgescu-Roegen are still unappreciated philosophers, in the old sense of the term.[81] Of course, "thermodynamics smacks of anthropomorphism" (as Max Planck lamented); but, as Georgescu-Roegen explained, "the idea that man can think of nature in wholly nonanthropomorphic terms is a patent contradiction in terms." Scale-dependent observation does not exist independent of the observer. It is particularly relevant in ecology, the science of complex systems ranging from the bacterial realm to the global Biosphere.

Like the Harvard physiologists formulating the homeostasis concept in the 1920-30s, Vernadsky was indebted to the great French physiologist Claude Bernard notably for his distinction (and exchange) between the "milieu intérieur" (internal environment) and the "milieu cosmique" (cosmic environment). Both Vernadsky and Lotka recognized Bernard's legacy. In his Sorbonne lectures, Vernadsky declared:

> In most of their works studying living organisms, the biologists disregard the indissoluble connection between the surrounding milieu and the living organism. In studying the organism as something quite distinct from the environment, the cosmic milieu, as Bernard said, they study not a natural body but a pure product of their thinking.

He pointed out a flagrant case of "fallacy of misplaced concreteness" (Alfred North Whitehead). Vernadsky immediately added: "For a long time the great biologists have seen the indissoluble connection between the organism and the surrounding milieu."[82]

[81] Georgescu-Roegen, 1971, p.276; complete bibliography in Georgescu-Roegen, 1995.

[82] Vernadsky, 1924 (p. 43). Quotation from Bernard, 1878, t. I, p.67. Bernard's theory of the constancy of the *milieu intérieur* was a decisive precedent in the discovery of physiological *homeostasis*, a term coined in 1926 (explained in 1929) by Harvard physiologist Walter Cannon (1871-1945), author of the Sorbonne lectures (Paris, 1930) published under the title *The Wisdom of the Body* (1932). The concept of homeostasis gained wide currency only after 1948 through the influence of Norbert Wiener (1894-1964), the father of cybernetics. Cannon was a close colleague of Lawrence J. Henderson (1878-1942), the author of *The Fitness of the Environment* (1913), an inspiring book quoted by Vernadsky, which belongs to the prehistory of Gaia. See Langley, 1973.

It is "living matter," Vernadsky explained, that, "prepared itself a new 'cosmic' milieu, taking this notion in Bernard's sense."[83] This reference is capital: it constitutes an important step in the story of the Gaian science of *geophysiology*. Vernadsky's approach may be compared with contemporary discussions about biological organization, the order of nature, the balance of nature's economy, self-regulating systems, processes of equilibrium, steady states, and geochemical cycles. The principles of thermodynamics helped to form a new cosmological viewpoint, a unifying framework for the study of natural processes, connecting living systems with the energetic economy of nature. Respiration and nutrition, as manifestations of energy-matter exchange between the organism and the environment, connect life with the cosmos. There is a natural affinity between Vernadsky's thermodynamic worldview and Lotka's concept of "world engine." Both Vernadsky and Lotka can be considered sources for Georgescu-Roegen's new paradigm of bioeconomics.[84] After Vernadsky, Lotka, Gaia theory, and Georgescu-Roegen, the world *problematique* of sustainable development can and must integrate bioeconomics and biogeochemistry, our global "industrial metabolism" and the matter-energy system of Earth's Biosphere.

There are interesting links between the Carnotian revolution, the Wegenerian revolution, and the Vernadskian revolution, between thermodynamics, dynamic Earth, biogeochemistry, and bioeconomics, following the revolutionary epistemological work of N. Georgescu-Roegen.[85] Our modern culture must integrate the entropy law, the mobilist view of the dynamic Earth and the Biosphere. A growing intellectual circle considers Vernadsky's *The Biosphere* as a classic of scientific thought on a level equal to Darwin's *Origins of Species*. In one of the fundamental environmental books of the early 1970s, the American ecologist Howard T. Odum, Hutchinson's former student in biogeochemistry, wrote:

> We can begin a systems view of the earth through the macroscope of the astronaut high above the earth. From an orbiting satellite, the earth's living zone appears to be very simple. The thin water- and air-bathed shell covering the earth—the biosphere—is bounded on the inside by dense solids and on the outside by the near vacuum of outer space.[86]

James Lovelock's concept of Gaia first appeared in this context: "The start of the Gaia hypothesis was the view of the Earth

[83] Bernard, 1878b, p.67.

[84] Grinevald, 1987; and Krishnan, 1995.

[85] Georgescu-Roegen, 1971. This is his magnum opus.

[86] Odum, 1971a, p. 11.

from space, revealing the planet as a whole but not in detail."[87] At the time, Vernadsky was forgotten in the West. Significantly, in his review (*New Scientist*, 17 July 1986) of the booklet *The Biosphere*, by V. Vernadsky, the very abridged English translation edited by Synergetic Press, Lovelock confessed:

> When Lynn Margulis and I introduced the Gaia hypothesis in 1972 neither of us was aware of Vernadsky's work and none of our much learned colleagues drew our attention to the lapse. We retraced his steps and it was not until the 1980s that we discovered him to be our most illustrious predecessor.

Now the time is ripe for the revival of the real historical figure of Vernadsky and his complete work on the Biosphere. We must realize that the natural system of Earth, named Nature by Humboldt, the Biosphere by Vernadsky, Gaia by Lovelock, and ecosphere by others is a fundamental concept for our religious, philosophical and scientific quest to learn "What Is Life?," as emphasized by Vernadsky himself and recently reformulated in an admirable book, fifty years after Schrödinger, by Lynn Margulis and Dorion Sagan.[88]

Nothing is as hard or as necessary as understanding Life. Our own individual life and our collective life is, essentially, the activity of the Biosphere —it is *creative evolution*. We are, as far as we know, the only naturally habitable planet in the solar system, and perhaps even—although we are completely ignorant on this scale—in the immense cosmos.

To paraphrase Michael Ruse's *Taking Darwin Seriously*, more than a half century after the appearance of *The Biosphere* the time has surely come to take Vernadsky seriously.

Jacques Grinevald
University of Geneva
Switzerland

[87] Lovelock, 1979, p.126. There is no reference to Vernadsky in the first scientific papers on the Gaia hypothesis published by Lovelock and Margulis in the 1970s. The references to Vernadsky appeared in the mid-1980s, notably in Lovelock, 1988. For a complete bibliography, see Margulis and Sagan, 1997.

[88] Margulis and Sagan, 1995.

Translator's Preface

Translation of *The Biosphere* has been materially aided by access to an English version, prepared some years ago, in a form that faithfully followed the French edition. The revision presented here is a rather drastic one, in which the sequence of ideas in sentences and paragraphs has been rearranged in the interests of compactness and logical flow. It may have been rash to take such liberties, since Vernadsky's German translator and the Russian editor of his posthumous book both were scrupulous in following the author literally. Both, however, felt it necessary to explain this and beg the reader's understanding. It is believed that the present version conscientiously preserves Vernadsky's meaning, and it is hoped that it will give the reader an interest in, and some understanding of, this extraordinary scientist and his work.

Both the French and Russian editions[89] were used in making this revised translation. In cases of doubt, the Russian version was deemed the more authoritative. The appendix to the French edition, on Evolution of Species, is not included here, nor is it in the 1967 Russian edition.

Dr. Richard Sandor's help with part of the translation is appreciated.

D. B. Langmuir
June 1977

[89] Editions of V. I. Vernadsky's book published in the past include the following. In Russian: *Biosfera [The Biosphere]*, 1926, Leningrad, Nauch. Kim-Tekhn. Izdatel'stvo; *Izbrannye Sochineniya [Selected Works]*, 1960, Moscow, Izdatel'stvo Akademiya Nauk SSSR, v. 5, pp. 7-102; *Biosfera [The Biosphere]*, 1967 [footnotes below ascribed to A. I. Perelman are translated from this edition by D. Langmuir], Mysl', Moscow, pp. 222-348; *Biosfera i Noosfera [The Biosphere and the Noösphere]*, 1989, Moscow, Nauka, 115 p.; *Zhivoe Veshchestvo i Biosfera [Living Matter and the Biosphere]*, 1994, Moscow, Nauka, pp. 315-401. In French: *La Biosphère*, 1929, Paris, F. Alcan, 232 p. In Serbo-Croatian: *Biosfera*, 1960, Beograd, Kultura, 233 p. In Italian: *La Biosfera*, 1993, Como, Italy, Red Edisioni. In English: only the an abridged version based on the French edition of 1929 has been available: *The Biosphere*, 1986, Oracle, Arizona, Synergetic Press.

Editor's Note on Translation and Transliteration

The revised translation you hold has been modified somewhat from David Langmuir's translation, which dates back to the 1970's. Langmuir's overall organization of the text was well done and has largely been retained. Conflicts between the French and Russian editions have been noted.

In this translation, the text has been rendered into more fluent English. In many places, this has involved adding words and phrases that will not be found in the original. This is not, however, a problem for the fidelity of the translation, for Vernadsky's sense has been completely retained. Much that is implicit in Russian writing must be made explicit in English for clarity. Doing so has dramatically improved the readability of this book for English-speaking audiences.

The most difficult word to translate in this book is the verb (and its derivatives) *ugadyvat´/ugadat´,* "to guess." In an authoritative dictionary of Russian verbs, the verb *ugadyvat´/ugadat´* is rendered simply as "guess." It is a verb whose 1st and 2nd person forms are not used in contemporary Russian (S. Rowland, personal communication), and E. Daum and W. Schenk define[90] it to mean to be guessed, to seem "when no definite information is available on the subject."

David Langmuir translated *ugadyvat´/ugadat´* as "prediction," Andrei V. Lapo as "guess," but I see neither of these as entirely adequate in this context. The word for "conjecture" in Russian is *dogadka* or *gadanie*, and both, as well as *ugadyvat´/ugadat´*, share the root "gad." The phrase for "scientific guess" is *nauchnoe predpolozhenie*, whereas the phrases for "by guess" or "lucky guess" are *naugad* and *schastlivaya dogadka*, respectively.[91] "Extrapolation" translates as the cognate *ekstrapolyatsiya*, whereas "guesswork" is rendered as *dogadki*. Thus words using the "gad" root seem to imply an element of informality. But the present context is clearly one of formal scientific research. Therefore, by using *ugadyvat´/ugadat´* in noun form Vernadsky apparently intends it to mean something along the lines of "conjectural constructs founded on

90 Daum and Schenk, 1974.

91 Gal'perin, 1972, p. 616.

guesswork," and there is no question that he uses it in a pejorative sense.

The exact meaning is important here, for it underpins Vernadsky's critique of certain mechanistic models of the world which have a tendency to construct risky extrapolations. It serves to highlight the contrast between the way science is conducted in the East and in the West. Vernadsky may, in fact, have been directing this criticism at the way he felt scientific research was being conducted in the west.[92] A clarification of this point may help us to understand both the differences between Russian and western scientific epistemology and the contrasts between Russian and western approaches to environmental problems. Perhaps ironically, Vernadsky is being increasingly identified in the west as a doyen of the environmental movement.[93]

Although he is surely not the most orthodox model of a western scientist, science fiction author Arthur C. Clarke in a recent interview vividly expressed the extrapolationist approach to science in the west. The interviewer, noting that some of Clarke's ideas (such as geostationary satellites and ice on the moon) had made the jump from science fiction to science fact, asked Clarke if he had hopes that any other of the predictions in his works of fiction would make the transition to confirmed fact. Clarke replied: "Actually, I very seldom predict, but I extrapolate, and there are many things I extrapolate that I would hate to be accurate predictions."[94]

Extrapolationistic representations are thus in direct conflict with Vernadsky's preference for advancing science through empirical generalizations—those observations that, once pointed out, are inescapable to any observer. Although empirical generalizations can change over time with new evidence or new interpretations, according to Vernadsky they should never involve conjecture, extrapolation or guesswork. They thus constitute Vernadsky's bedrock starting point for scientific investigation. As you will see, however, there are places where Vernadsky himself departs from this idealized course of action.[95]

As pointed out by Andrei Lapo, the translation of another word is critical for understanding this book. This is the Russian word *kosnoe*, an antonym of "living." Langmuir usually translated this word as "crude" (as an English equivalent to the French "brut"), and sometimes translated it as "inert." George Vernadsky (son of Vladimir Vernadsky) always, however, trans-

[92] Although the criticism could surely also be applied to some of his Russian colleagues.

[93] Margulis and Sagan, 1985; Matias and Habberjam, 1984; and Meyer, 1996.

[94] Guterl, 1997, pp. 68-69.

[95] Vernadsky did on occasion employ the concept of a "working scientific hypothesis." See p. 10 of his "On the geological envelopes of Earth as a planet," paper read before the meeting of the Sechnov Institute of Scientific Research, January 18, 1942.

lated *kosnoe* as "inert." Both Lapo and I are in agreement with G. Vernadsky that "inert" is the best translation for *kosnoe*. Translation of this word is important because Vernadsky divided the biosphere into two classes of materials, the actual living matter and natural, bio-inert matter with which it is associated.

Dr. Lapo also noted that there are places in this book where it is not perfectly clear whether V. Vernadsky meant to refer to the "surrounding medium" or to the "environment" in a more general sense. In reviewing the translation of V. Vernadsky's words, I have tried to maintain a faithful usage of both. This distinction is important because in his view of the biosphere Vernadsky emphasizes, with a nod to the work of L. Pasteur,[96] the "medium forming" capabilities of organisms.

Mark McMenamin
Department of Geology and Geography
Mount Holyoke College
South Hadley, Massachusetts
May 1997

[96] Vernadsky apparently took tremendous inspiration from Louis Pasteur. There are parallels between the careers of both scientists. Pasteur's first scientific breakthrough was in the field of crystallography. In 1848 he founded the discipline of stereochemistry with his discovery that tartaric acid (isolated from fruits) came in three varieties, laevo-tartarate, dextro-tartarate and a racemic mixture (paratartaric acid) of the other two. Following Eilhard Mitscherlich's demonstration that the three compounds are chemically identical, Pasteur solved the problem of the opposite rotation of polarized light by laevo-tartarate and dextro-tartarate by elegantly showing (after hand-picking the tiny crystals) that the two substances formed mirror-image crystalline forms (Vallery-Rodot, 1912; and Compton, 1932.).

When young Vernadsky was in Paris in 1889, studying with Henri Louis Le Châtelier and Ferdinand André Fouqué, he carried out experimental studies on silicate minerals and succeeded for the first time in producing synthetic sillimanite (Shakhovskaya, 1988, p.37; Eliseev and Shafranovskii, 1989; and Kolchinskii, 1987, pp. 11-12). An unexpected result of this artificial synthesis was Vernadsky's recognition of two different crystalline forms of sillimanite. This sillimanite polymorphism led Vernadsky to consider the problem of polymorphism in general for his *magister* dissertation. This work was of great geological importance, for "Vernadsky demonstrated the presence of a fundamental radical in most aluminosilicates, thereby uniting nearly all silicates into a unified system" (Rowland, 1993, p. 246). For his doctoral dissertation, Vernadsky addressed the phenomenon of gliding in crystals, and published his first major scientific work (Vernadsky, 1903).

THE BIOSPHERE

Author's Preface to the French Edition

After this book appeared in Russian in 1926, the French edition
was revised and recast to correspond with the Russian text. It
forms the continuation of the essay *La Géochimie*, published in
the same collection (1924), of which a Russian translation has
just appeared and a German translation is to appear shortly.[1]

Only a few bibliographical references will be found here; for
others please consult *La Géochimie*. The problems discussed
herein have been touched upon in several articles, most impor-
tantly in the "Revue Generale des Sciences" (1922-1925), and in
"Bulletins of the Academy of Sciences of the USSR in Leningrad"
(St. Petersburg) (1826-1927), both in French.

The aim of this book is to draw the attention of naturalists,
geologists, and above all biologists to the importance of a quantitative study of the relationships between life and the chemical
phenomena of the planet.

Endeavoring to remain firmly on empirical grounds, without
resorting to hypotheses, I have been limited to the scant number
of precise observations and experiments at my disposal. A great
number of quantitatively expressed empirical facts need to be
collected as rapidly as possible. This should not take long once
the great importance of living phenomena in the biosphere
becomes clear.

The aim of this book is to draw attention to this matter; hopefully it will not pass unnoticed.

V. I. Vernadsky
December 1928

1 Vernadsky, 1930, p. 370.

Author's Preface to the Russian Edition

Among numerous works on geology, none has adequately treated the biosphere as a whole, and none has viewed it, as it will be viewed here, as a single orderly manifestation of the mechanism of the uppermost region of the planet — the Earth's crust.

Few scientists have realized that the biosphere is subject to fixed laws. Since life on Earth is viewed as an accidental phenomenon, current scientific thought fails to appreciate the influence of life at every step in terrestrial processes. Earth scientists have assumed that there is no relationship between the development of life on Earth and the formation of the biosphere — the envelope of life where the planet meets the cosmic milieu.

Historically, geology has been viewed as a collection of events derived from insignificant causes, *a string of accidents*. This of course ignores the scientific idea that geological events are *planetary phenomena*, and that the laws governing these events are not peculiar to the Earth alone. As traditionally practiced, geology loses sight of the idea that the Earth's structure is a harmonious integration of parts that must be studied as an indivisible mechanism.

As a rule, geology studies only the details of phenomena connected with life; the *mechanism* behind the details is not regarded as a scientific phenomenon. Though surrounded by manifestations of such mechanisms, insufficiently attentive investigators frequently overlook them.

In these essays, I have attempted to give a different view of the geological importance of living phenomena. I will construct no hypotheses and will strive to remain on the solid ground of empirical generalization, thus portraying the geological manifestations of life and planetary processes from a base of precise and incontestable facts.

I will bypass, however, three preconceived ideas, long a part of geological thought, that seem to contradict the method of empirical generalization in science, generalizations which constitute the fundamental discoveries of natural scientists.

One of these conceptions, mentioned above, is that geological phenomena are *accidental coincidences of causes*, essentially blind, and obscure because of their complexity and number.

This preconceived idea[2] is only partly related to certain philosophical-religious interpretations of the world; mainly it is based upon incomplete logical analysis of our empirical knowledge.

Two other preconceived ideas have infiltrated geology from roots foreign to the empirical principles of science. The first is the assumption of the *existence of a beginning of life* — life genesis or biopoesis at a certain stage in the geological past. Considered a logical necessity, this has penetrated science in the form of religious and philosophical speculation.[3] Next is the preconception of an essential role for *pre-geological stages of planetary evolution* during which conditions were clearly different from those which can now be studied. In particular, an igneous-liquid or incandescent gaseous stage is assumed as certain. These ideas diffused into geology during early developments of philosophy and cosmogony.

The logical consequences of these ideas are illusory, harmful, and even dangerous when applied to contemporary geology.

Let us consider all empirical facts from the point of view of a *holistic mechanism* that combines all parts of the planet in an indivisible whole. Only then will we be able to perceive the perfect correspondence between this idea and the geological effects of life. I will not speculate here about the existence of the mechanism, but rather will observe[4] that it corresponds to all the empirical facts and follows from scientific analysis. Included in the mechanism as its integrating member is the biosphere, the domain of manifestation of life.

I have found no empirical evidence whatsoever for assuming a beginning of life, nor for the influence of cosmic planetary states on geological events, nor for the existence of an early igneous state, and I regard these concepts as useless and restricting to valid scientific generalizations. Philosophical and cosmogonal hypotheses that cannot be founded on facts should be discarded, and replacements sought.

The Biosphere in the Cosmos and *The Domain of Life*, the two essays that make up this book, are independent works linked by the point of view set forth above. The need for writing them became apparent as a result of studies of the biosphere conducted by the author since 1917.

In connection with this work I have written three essays, as follows: "Living Matter," "The Structure of Living Matter," and

2 Compare this with the "ruling theory" concept of T. C. Chamberlain (1965). A ruling theory is one which skews the ability of the researcher (holding the ruling theory) to objectively evaluate new data. Often this occurs because the researcher has come up with the idea his—or herself, and out of pride of parenthood would not wish to see the theory threatened by nonconforming data. Such data thus tend to be inappropriately discounted by victims of the ruling theory syndrome.

3 Here Vernadsky has developed a slavic version of systematic or substantive uniformitarianism, the theory that nothing on Earth really ever changes (not to be confused with actualism or methodological uniformitarianism, the research technique used by all geologists in which they use present phenomena as the key to past processes; see A. Hallam, 1992). There is an interesting parallel here with western geological thought. The actualistic principles of J. Hutton (1795) were overextended to the notion that Earth "has no stratigraphy whatsoever, and hence no real history; it is instead a model of a system in dynamic equilibrium" (Hallam, 1992, p. 25). This substantive uniformitarianism was elaborated and (over)extended by C. Lyell (1830-1833). Lyell's view was lampooned by his colleague Henry de la Beche in a cartoon showing a future "Prof. Ichthyosaurus" lecturing to fellow aquatic reptiles about the odd appearance of the extinct human animal now known only by its skull. Indeed, Lyell felt that when appropriate conditions of climate and temperature reappear, the dinosaurs, other extinct animals and extinct vegetation will return.

Vernadsky received actualistic principles from the tradition of the great Russian scientist M. V. Lomonosov (1711-1765), who was one of the first scientists to apply actualism to geological problems (Tikhomirov, 1969; and Vernadsky, 1988, pp. 326-328). Andrei Lapo calls Vernadsky the "Lomonosov of the Twentieth Century" (Lapo, 1988, pp. 3-10).

Vernadsky's dismissal of "the *existence of a beginning of life*" (emphasis his) is testimony to his allegiance to the principle of actualism (Kurkin, 1989, pp. 516-528) and

"Living Matter in Geochemiscal History of the Element's System." I have not had time to prepare these for publication, but hope to do so at a later date.[5]

Vladimir Vernadsky
Prague
February 1926

is also the Russian counterpart to western substantitive uniformitarianism. Vernadsky sees Earth as a planet on which life has always been present. Life is thus an essential criterion for characterizing Earth as a planet. For Vernadsky, discussion of the origin of life on Earth is not within the realm of science (let alone geology) and is merely an antiquated form of "religious and philosophical speculation." Here then is the parallel with Hutton: to paraphrase the famous last line of Hutton's book, for Vernadsky life has "no vestige of a beginning, no prospect of an end." Also implied here is a harsh critique of the Oparin concept, first dating from the early 1920's, of biopoesis (abiotic genesis of life) in an early Earth's reducing atmosphere. Vernadsky did not categorically deny the possibility of biopoesis, but insisted that it could not occur during the course of *geological* time known to us.

Vernadsky rejects the idea of the origin of life in an early reducing atmosphere. This helps explain why Vernadsky missed the greatest research insight available to his research program, the discovery of the Oxygen Revolution at two billion years ago. This discovery fell to the American geologist Preston Cloud (1973). Vernadsky would have appreciated the significance of Cloud's discovery, for Vernadsky was a student and an advisee of the great soil scientist and teacher V. V. Dokuchaev, and the pervasive oxidation so characteristic of both modern and ancient soils does not appear in the geological record until after the Oxygen Revolution. It was Dokuchaev, Vernadsky wrote in 1935, who "first turned my attention to the dynamic side of mineralogy" (cited in

Bailes, 1990, p. 19) and "put forward his thesis that the soil is a peculiar natural body, different from rock" (Vernadsky, 1944, p. 490). Bailes (p. 19) noted how Dokuchaev fired up "Vernadsky's interest in a holistic and historical approach to science; indeed, Vernadsky held the history of science on a par with other types of scientific investigation (Mikulinsky, 1984). Unlike western substantive uniformitarians, Vernadsky does not deny Earth a history. On the contrary, he has an abiding sense of the history of the planet, as in this quotation by Vernadsky on p. 42 of Bailes' book: "minerals are remains of those chemical reactions which took place at various points on Earth; these reactions take place according to laws which are known to us, but which, we are allowed to think, are closely tied to general changes which Earth has undergone as a planet. The task is to connect the various phases of changes undergone by the earth with the general laws of celestial mechanics." Vernadsky (Bailes, p. 83) also wrote of "the beauty of historical phenomena, their originality among the other natural processes."

Vernadsky eventually came to a reluctant acceptance of abiogenesis (development of life from non-life; discussed mainly in his posthumous publications), and in fact championed the idea of multiple "bioclades" (a term coined by Raup and Valentine, 1983) or, in other words, the polyphyletic origin of life. See Rogal' (1989) and Borkin (1983). In 1908 Vernadsky championed the hypothesis of directed panspermia, perhaps because of its possible bearing on his concept of the eternity of life.

"By the way, it turns out that the quantity of living matter in the earth's crust is a constant. Then *life* is the same kind of part of the cosmos as energy and matter. In essence, don't all the speculations about the arrival of 'germs' [of life] from other heavenly bodies have basically the same assumptions as [the idea of] the eternity of life?" (p. 123, Bailes)

Bailes (p. 123) criticizes this passage (taken from a letter to Vernadsky's former student I. V. Samoilov) as being "rather cryptic," "not very clear," and as evidence of a "mystical strain" in Vernadsky's thought. Similar criticisms were also levied by Oparin (1957). These criticisms are misleading, however; Vernadsky saw life's development, and its subsequent development of the noösphere, as a materialistic process. Vernadsky eschewed vitalism (see Vernadsky, 1944, p. 509):

New vitalistic notions have their foundation not in scientific data, which are used rather as illustrations, but in philosophical concepts such as Driesch's "entelechy."[Driesch, 1914] The notion of a peculiar vital energy (W. Ostwald) is likewise connected with philosophical thought rather than with scientific data. Facts did not confirm its real existence.

Vernadsky probably would have agreed with the statement of Humberto R. Maturana and Francisco J. Varela that living "systems, as physical autopoietic machines, are purposeless systems" (1980). Vernadsky nevertheless permitted the noösphere to form as a logical outcome of these purposeless machines.

4 Here Vernadsky asserts that he is worthy of making empirical generalizations.

5 Only one of these essays has been published to date (Vernadsky, 1978).

PART ONE

The Biosphere[6] in the Cosmos

In nature all is harmony,
A consonance fore'er agreed on....

F. Tyutchev, 1865

The Biosphere in the Cosmic Medium

1 The face of the Earth[7] viewed from celestial space presents a unique appearance, different from all other heavenly bodies. The surface that separates the planet from the cosmic medium is the *biosphere*, visible principally because of light from the sun, although it also receives an infinite number of other radiations from space, of which only a small fraction are visible to us. We hardly realize the variety and importance of these rays, which cover a huge range of wavelengths.

Our understanding is full of gaps, but improved detectors are rapidly expanding our knowledge of their existence and variety. Certainly they make the empty cosmic regions different from the ideal space of geometry![8]

Radiations reveal material bodies and changes in the cosmic medium. One portion appears as energy through transitions of states, and signals the movements of aggregates of quanta, electrons and charges. The aggregates, which as a whole may remain motionless, control the movements of their separate elements.

There are also rays of particles (the most-studied are electrons) which often travel at nearly the same speed as waves, and result from transitions in separate elements of the aggregates. Both kinds of rays are powerful forms of energy, and cause observable changes when they pass through material bodies.

2 For the moment, we can neglect the influence of particle radiation on geochemical phenomena in the biosphere, but we must always consider the radiations from transitions of energy states. These will appear as light, heat, or electricity according to their type and wavelength, and produce transformations in our planet.

These rays cover a known range of forty octaves in wavelength (10^{-8} cm to kilometers), of which the visible spectrum is one octave.[9] This immense range is constantly being extended by scientific discovery, but only a few of the forty octaves have thus far affected our view of the cosmos.

The radiations that reach our planet from the cosmos amount to only four and one-half octaves. We explain the absence of the other octaves on the Earth's surface by absorption in the upper atmosphere.

The best-known radiations come from the sun — one octave of light rays, three of infrared radiation, and a half-octave of ultraviolet; the last half-octave being, doubtless, only a small fraction of the total ultraviolet from the sun, most of which is retained by the stratosphere. (§115)

6 "Biospherology" is the term now used by some for study of the biosphere (see Guegamian, 1980). Others, such as NASA, use the term "biospherics."

7 In this first phrase Vernadsky echoes the title and opening sentence of Eduard Suess's influential geological compendium, *Die Antlitz der Erde* [*The Face of the Earth*] (Suess, 1883-1909, p. 1). Suess wrote:

"If we imagine an observer to approach our planet from outer space, and, pushing aside the belts of red-brown clouds which obscure our atmosphere, to gaze for a whole day on the surface of the earth as it rotates beneath him, the feature beyond all others most likely to arrest his attention would be the wedge-like outline of the continents as they narrow away to the South."

For more information on Suess's influence, see Greene (1982). Vernadsky admired the inductive approach utilized by Suess in this book.

8 Vernadsky reached the conclusion early on that radiation from the cosmos played a large role in the development of life.

9 "Octave" is a term used in both music and physical science. It means the same thing in both: a span over which a wavelength is halved or doubled.

3 A new character is imparted to the planet by this powerful cosmic force. The radiations that pour upon the Earth cause the biosphere to take on properties unknown to lifeless planetary surfaces, and thus transform the face of the Earth. Activated by radiation, the matter of the biosphere collects and redistributes solar energy, and converts it ultimately into free energy capable of doing work on Earth.

The outer layer of the Earth must, therefore, not be considered as a region of matter alone, but also as a region of energy and a source of transformation of the planet. To a great extent, exogenous cosmic forces shape the face of the Earth, and as a result, the biosphere differs historically from other parts of the planet. This biosphere plays an extraordinary planetary role.

The biosphere is at least as much a *creation of the sun* as a result of terrestrial processes. Ancient religious intuitions that considered terrestrial creatures, especially man, to be *children of the sun* were far nearer the truth than is thought by those who see earthly beings simply as ephemeral creations arising from blind and accidental interplay of matter and forces. Creatures on Earth are the fruit of extended, complex processes, and are an essential part of a harmonious[10] cosmic mechanism, in which it is known that fixed laws apply and chance does not exist.[11]

4 We arrive at this conclusion via our understanding of the matter of the biosphere — an understanding that had been profoundly modified by contemporary evidence that this matter is the direct manifestation of cosmic forces acting upon the Earth.

This is not a consequence of the extraterrestrial origin of matter in the biosphere, perhaps the majority of which has fallen from space as cosmic dust and meteorites. This foreign matter cannot be distinguished in atomic structure from ordinary terrestrial matter.

We must pause before entering the domain of terrestrial phenomena, because our ideas about the unforeseen character of matter on this planet are going through great transformations, upsetting our understanding of geology.

The identity of structure[12] between earthly matter and exogenic cosmic matter is not limited to the biosphere, but extends through the whole terrestrial crust; i.e., through the lithosphere, which extends to a depth of 60-100 kilometers, and interfaces with the biosphere at its outermost part. (§89)

Matter in the deeper parts of the planet shows the same identity, although it may have a different chemical composition.

10 Cf. the Tyutchev epigraph above.

11 Vernadsky is quite explicit here in his challenge to the "randomness" component of materialistic darwinism. This component has been expressed, complete with reference to the biosphere, by J. Monod (1971, p. 98):
"Randomness caught on the wing, preserved, reproduced by the machinery of invariance and thus converted into order, rule, necessity. A *totally* blind process can by definition lead to anything; it can even lead to vision itself. In the ontogenesis of a functional protein are reflected the origin and descent of the whole biosphere."

12 Presumably Vernadsky here means identity of atomic structure.

Matter from these regions seems, however, not to penetrate to the Earth's crust even in small amounts, and can therefore be ignored in studies of the biosphere.[13]

5 The chemical composition of the crust has long been regarded as the result of purely geological causes. Explanations of it have been sought by invoking the action of waters (chemical and solvent), of the atmosphere, of organisms, of volcanic eruptions, and so on, assuming that geological processes and the properties of chemical elements have remained unchanged.[14]

Such explanations presented difficulties, as did other and more complicated ideas that had been proposed. The composition was considered to be the remains of ancient periods when the Earth differed greatly from its present state. The crust was regarded as a scoria formed on the terrestrial surface from the once-molten mass of the planet, in accordance with the chemical laws that apply when molten masses cool and solidify.[15] To explain the predominance of lighter elements, reference was made to cosmic periods before the formation of the crust. It was thought that heavier elements were collected near the center of the Earth, during its formation as a molten mass thrown off from a nebula.

In all these theories, the composition of the crust was seen as a result of strictly geological phenomena. Chemical changes in composition of the crust were attributed to geological processes acting at lower temperatures, whereas isotopic changes in crustal composition were attributed to processes acting at higher temperatures.

6 These explanations are decisively contradicted by newly established laws which are in accord with recent results indicating that the chemical composition of stars is marked by previously unsuspected complexity, diversity, and regularity.[16]

The composition of the Earth, and particularly its crust, has implications that transcend purely geological phenomena. To understand them, we must direct our attention to the composition of all cosmic matter and to modifications of atoms in cosmic processes. New concepts are accumulating rapidly in this speculative field. Comparatively little theoretical analysis has been done, however, and deductions that might be justifiable have seldom been made explicit. The immense importance and unexpected consequences of these phenomena cannot, however, be disregarded. Three aspects of these phenomena can be

13 Study of deep seated kimberlite pipe eruptions (mantle rocks that somehow penetrated to Earth's surface) demonstrate that this can no longer be strictly the case; see Nixon, 1973; and Cox, 1978.

14 These are the fundamental assumptions of geological actualism.

15 Such an assumption led Lord Kelvin (see Hallam, 1992, p. 124; and Kelvin, 1894) to an erroneous calculation of the age of Earth.

16 Here Vernadsky is without doubt referring to the work of Einar Hertzsprung and Henry Norris Russell. Hertzsprung's pioneering research advanced the knowledge concerning the color of stars. Star color can be used as an index to star temperature. Russell's work greatly extended the list of stars with known luminosities (as calculated by parallax measurements). Plots of stellar luminosity to surface temperature, published by Russell beginning around 1915, established the "main sequence" of stars in the universe. Using the theoretical Hertzsprung-Russell diagram, one may plot lines of constant stellar radius against an ordinate axis of luminosity and an abcissa axis of effective temperature. Thus if one knows the luminosity and effective temperature of a star, it is possible to calculate its radius.

It became possible to remotely analyse the composition of stars when the lines in the solar spectrum (named Fraunhofer lines after the glass maker and optician Joseph Fraunhofer) were explained by photographer W. H. Fox Talbot (1800-1877) and Gustav Kirchoff (1824-1887) as absorption lines characteristic (with absorption occuring as sunlight passes through the cooler, outer gaseous layers of the Sun) for specific excitation states of particular elements.

given preliminary discussion, namely: 1. the peculiar positions of the elements of the crust in Mendeleev's periodic system;[17] 2. their complexity; 3. the non-uniformity of their distribution.

Elements with even atomic number clearly predominate in the Earth's crust.[18] We cannot explain this by known geological causes. Moreover, the same phenomenon is more marked in meteorites, the only bodies foreign to the Earth that are immediately accessible for study.[19]

The two other aspects seem even more obscure. The attempts to explain them by geological laws or causes apparently contradict well-known facts. We cannot understand the hard facts of the complexity of terrestrial elements; and still less, their fixed isotopic compositions. Isotopic ratios in various meteorites have been shown to be the same,[20] in spite of great differences in the history and provenance of these meteorites.

Contrary to previous beliefs, it is becoming impossible to perceive the laws that govern the Earth's composition in terms of purely geological phenomena, or merely in terms of "stages" in the Earth's history. The latter explanation fails on account of the fact that there is neither a similarity of the deeper portions of our planet with the composition of meteorites, nor, as in meteorites,[21] an even mix of both lighter chemical elements and of denser iron in rocks of either Earth's crust or rocks from depth. The hypothesis that elements will be distributed according to weight, with the heaviest accumulating near the center, during the formation of the Earth from a nebula, does not agree with the facts. The explanation can be found neither in geological and chemical phenomena alone, nor in the history of the Earth considered in isolation. The roots lie deeper, and must be sought in the history of the cosmos, and perhaps in the structure of chemical elements.[22]

This view of the problem has recently been confirmed, in a new and unexpected way, by the similarity in composition between the Earth's crust and the sun and stars. The likeness in composition of the crust and the outer portions of the sun was noted by Russell as early as 1914, and the resemblances have become more marked in the latest work on stellar spectra.[23] Cecilia H. Payne[24] lists heavier stellar elements in descending order of abundance as follows: silicon, sodium, magnesium, aluminum, carbon, calcium, iron (more than one percent); zinc, titanium, manganese, chromium, potassium (more than one per mil).

This pattern clearly resembles the order of abundance in the

17 See Mendeleev, 1897.

18 See Oddo, 1914.

19 See Harkins, 1917.

20 This observation was later used to date all of the meteorites (and Earth itself) to age of approximately 4.6 billion years based on abundances of Strontium-87 and Rubidium-87 (Reynolds, 1960).

21 See Farrington, 1901.

22 Vernadsky was overinterpreting his data here. Iron and nickel went to Earth's core at a time when the planet was completely or partially melted, early in its history. The flow of dense liquid toward the core released additional heat (as thermal [kinetic] energy converted from potential energy) and caused additional melting of rock.

23 See Norris, 1919.

24 See Payne, 1925.

Earth's crust: oxygen, silicon, aluminum, iron, calcium, sodium, potassium, magnesium.

These results, from a new field of study, show striking similarities between the chemical compositions of profoundly different celestial bodies. This might be explained by a material exchange taking place between the outer parts of the Earth, sun, and stars. The deeper portions present another picture, since the composition of meteorites and of the Earth's interior is clearly different from that of the outer terrestrial envelope.

7 We thus see great changes occurring in our understanding of the composition of the Earth, and particularly of the biosphere. We perceive not simply a planetary or terrestrial phenomenon, but a manifestation of the structure, distribution, and evolution of atoms throughout cosmic history.

We cannot explain these phenomena, but at least we have found that the way to proceed is through a new domain of phenomena, different from that to which terrestrial chemistry has so long been limited. Viewing the observed facts differently, we know *where* we must seek the solution of the problem, and where the search will be useless. The structure of the cosmos manifests itself in the outer skin or upper structure of our planet. We can gain insight into the biosphere only by considering the obvious bond that unites it to the entire cosmic mechanism.[25]

We find evidence of this bond in numerous facts of history.

The Biosphere as a Region of Transformation of Cosmic Energy

8 The biosphere may be regarded as a region of transformers that convert cosmic radiations into active energy in electrical, chemical, mechanical, thermal, and other forms. Radiations from all stars enter the biosphere, but we catch and perceive only an insignificant part of the total; this comes almost exclusively from the sun.[26] The existence of radiation originating in the most distant regions of the cosmos cannot be doubted. Stars and nebulae are constantly emitting specific radiations, and everything suggests that the penetrating radiation discovered in the upper regions of the atmosphere by Hess[27] originates beyond the limits of the solar system, perhaps in the Milky Way, in nebulae, or in stars of the Mira Ceti type.[28] The importance of this will not be clear for some time,[29] but this penetrating cosmic radiation determines the character and mechanism of the biosphere.

[25] This view is also developed in the works of Alexandr E. Fersman (1933, 1934, 1937 and 1939). Fersman, Vernadsky's most influential student (Vernadsky, 1985; and Fersman, 1945), outlived his mentor by only a few months (Backlund, 1945). Fersman's work is not surprisingly an extension of the Vernadskian research program (Saukov, 1950).

[26] And of that we receive only one half billionth of the total solar output (Lovins, Lovins, Krause, and Bach, 1981).

[27] See Hess, 1928.

[28] Mira Ceti is a long period variable star. Variable stars show periodic variations in brightness and surface temperature. Mira Ceti has an average period of 331 days.

[29] Here Vernadsky anticipates the discovery of cosmic background radiation (Weinberg, 1988).

The action of solar radiation on earth-processes provides a precise basis for viewing the biosphere as both a terrestrial and a cosmic mechanism. The sun has completely transformed the face of the Earth by penetrating the biosphere, which has changed the history and destiny of our planet by converting rays from the sun into new and varied forms of energy. At the same time, the biosphere is largely the product of this radiation.

The important roles played by ultraviolet, infrared, and visible wavelengths are now well-recognized. We can also identify the parts of the biosphere that transform these three systems of solar vibration, but the mechanism of this transformation presents a challenge which our minds have only begun to comprehend. The mechanism is disguised in an infinite variety of natural colors, forms and movements, of which we, ourselves, form an integral part. It has taken thousands of centuries for human thought to discern the outlines of a single and complete mechanism in the apparently chaotic appearance of nature.

9 In some parts of the biosphere, all three systems of solar radiation are transformed simultaneously; in other parts, the process may lie predominantly in a single spectral region. The transforming apparatuses, which are always natural bodies, are absolutely different in the cases of ultraviolet, visible and thermal rays.

Some of the ultraviolet solar radiation is entirely absorbed,[30] and some partly absorbed, in the rarefied upper regions of the atmosphere; i.e., in the stratosphere, and perhaps in the "free atmosphere", which is still higher and poorer in atoms. The stoppage or "absorption" of short waves by the atmosphere is related to the transformation of their energy. Ultraviolet radiation in these regions causes changes in electromagnetic fields, the decomposition of molecules, various ionization phenomena, and the creation of new molecules and compounds. Radiant energy is transformed, on the one hand, into various magnetic and electrical effects; and on the other, into remarkable chemical, molecular, and atomic processes. We observe these in the form of the aurora borealis, lightning, zodiacal light, the luminosity that provides the principal illumination of the sky on dark nights, luminous clouds, and other upper-atmospheric phenomena. This mysterious world of radioactive, electric, magnetic, chemical, and spectroscopic phenomena is constantly moving and is unimaginably diverse.

These phenomena are not the result of solar ultraviolet radia-

30 By ozone.

tion alone. More complicated processes are also involved. All forms of radiant solar energy outside of the four and one-half octaves that penetrate the biosphere (§2) are "retained"; i.e., transformed into new terrestrial phenomena. In all probability this is also true of new sources of energy, such as the powerful torrents of particles (including electrons) emitted by the sun, and of the material particles, cosmic dust, and gaseous bodies attracted to the Earth by gravity.[31] The role of these phenomena in the Earth's history is beginning to be recognized.

They are also important for another form of energy transformation — living matter. Wavelengths of 180-200 nanometers are fatal[32] to all forms of life, destroying every organism, though shorter or longer waves do no damage. The stratosphere retains all of these destructive waves, and in so doing protects the lower layers of the Earth's surface, the region of life.

The characteristic absorption of this radiation is related to the presence of ozone (the ozone screen (§115), formed from free oxygen — itself a product of life).

10 While recognition of the importance of ultraviolet radiation is just beginning, the role of radiant solar heat or infrared radiation has long been known, and calls for special attention in studies of the influence of the sun on geologic and geochemical processes. The importance of radiant solar heat for the existence of life is incontestable; so, too, is the transformation of the sun's thermal radiation into mechanical, molecular (evaporation, plant transpiration, etc.), and chemical energy. The effects are apparent everywhere — in the life of organisms, the movement and activity of winds and ocean currents, the waves and surf of the sea, the destruction of rock and the action of glaciers, the formation and movements of rivers, and the colossal work of snow and rainfall.

Less fully appreciated is the role that the liquid and gaseous portions of the biosphere play as accumulators and distributors of heat. The atmosphere, the sea, lakes, rivers, rain, and snow actively participate in these processes. The world's ocean acts as a heat regulator,[33] making itself felt in the ceaseless change of climate and seasons, living processes, and countless surface phenomena. The special thermal properties of water,[34] as determined by its molecular character, enable the ocean to play such an important role in the heat budget of the planet.

The ocean takes up warmth quickly because of its great specific heat, but gives up its accumulated heat slowly because of

[31] Earth's magnetic field actually plays a more important role in these phenomena, as demonstrated by the Van Allen Radiation Belts (see Manahan, 1994, p. 287, fig. 9.9).

[32] Certain bacteria can survive such irradiation.

[33] In other words, a maritime influence greatly moderates climate on land.

[34] Particularly its high heat capacity.

feeble thermal conductivity.[35] It transforms the heat absorbed from radiation into molecular energy by evaporation, into chemical energy through the living matter which permeates it, and into mechanical energy by waves and ocean currents. The heating and cooling of rivers, air masses, and other meteorological phenomena are of analogous force and scale.

11 The biosphere's essential sources of energy do not lie in the ultraviolet and infrared spectral regions, which have only an indirect action on its chemical processes. It is *living matter* — the Earth's sum total of living organisms — that transforms the radiant energy of the sun into the active chemical energy of the biosphere.

Living matter creates innumerable new chemical compounds by photosynthesis, and extends the biosphere at incredible speed as a thick layer of new molecular systems. These compounds are rich in free energy in the thermodynamic field of the biosphere. Many of the compounds, however, are unstable, and are continuously converted to more stable forms.

These kinds of transformers contrast sharply with terrestrial matter, which is within the field of transformation of short and long solar rays through a fundamentally different mechanism. The transformation of ultraviolet *and infrared* radiation takes place by action on atomic and molecular substances that were created entirely independently of the radiation itself. Photosynthesis, on the other hand, proceeds by means of complicated, specific mechanisms *created by photosynthesis itself*. Note, however, that photosynthesis can proceed only if ultraviolet[36] and infrared[37] processes are occurring simultaneously, transforming the energy in these wavelengths into active terrestrial energy.

Living organisms are distinct from all other atomic, ionic, or molecular systems in the Earth's crust, both within and outside the biosphere. The structures of living organisms are analogous to those of inert matter, only more complex. Due to the changes that living organisms effect on the chemical processes of the biosphere, however, living structures must not be considered simply as agglomerations of inert stuff. Their energetic character, as manifested in multiplication, cannot be compared geochemically with the static chemistry of the molecular structures of which inert (and *once*-living) matter are composed.

While the chemical mechanisms of living matter are still unknown, it is now clear that photosynthesis, regarded as an

35 Vernadsky's physics is mistaken here. Thermal conductivity will govern both heat uptake and release.

36 Indeed, vitamin synthesis can depend on ultraviolet irridation. The sterol ergosterol (from ergot fungus, yeast), similar to cholesterol, is a precursor of vitamin D_2. Upon ultraviolet irradiation of ergosterol at a frequency of 282 nanometers, ergosterol is converted to *cis*-trachysterol. With further irradiation, *cis*-trachysterol is converted into calciferol (vitamin D_2). When a cholesterol derivative 7-dehydrocholesterol (5,7-cholestadiene-3b-ol) is irradiated, it forms vitamin D_3, an even more potent form of the D vitamin. D vitamins can be considered to have a considerable biogeochemical importance, as they are required for the regulation of deposition of skeletal and dental calcium (Brown, 1975).

37 To maintain temperatures at which photosynthesis can occur.

energetic phenomenon in living matter, takes place in a particular chemical environment, and also within a thermodynamic field that differs from that of the biosphere's. Compounds that are stable within the thermodynamic field of living matter become unstable when, following death of the organism, they enter the thermodynamic field of the biosphere[38] and become a source of free energy.*

The Empirical Generalization and the Hypothesis

12 An understanding of the energetic phenomena of life, as observed in a geochemical context, provides proper explanation for the observed facts, as outlined above. But considerable uncertainties exist, on account of the state of our biological knowledge relative to our knowledge of *inert matter*. In the physical sciences, we have been forced to abandon ideas, long thought to be correct, concerning the biosphere and the composition of the crust. We have also had to reject long established, but purely geologic explanations (§6). Concepts that appeared to be logically and scientifically necessary have proved to be illusory. Correcting these misconceptions has had entirely unexpected effects upon our understanding of the phenomena in question.

The study of life faces even greater difficulties, because, more than in any other branch of the sciences, the fundamental principles have been permeated with philosophical and religious concepts alien to science.[39] The queries and conclusions of philosophy and religion are constantly encountered in ideas about the living organism. Conclusions of the most careful naturalists in this area have been influenced, for centuries, by the inclusion of cosmological concepts that, by their very nature, are foreign to science. (It should be added that this in no way makes these cosmological concepts less valuable or less profound.) As a consequence, it has become extremely difficult to study the big questions of biology and, at the same time, to hold to scientific methods of investigation practiced in other fields.

13 The vitalistic and mechanistic representations of life are two reflections of related philosophical and religious ideas that are not deductions based upon scientific facts.[40] These representations hinder the study of vital phenomena, and upset empirical generalizations.

Vitalistic representations give explanations of living phenomena that are foreign to the world of models — scientific general-

38 Here Vernadsky is making a very clear distinction between living matter and the non-living matter of the biosphere. This may be compared to the Treviranian concept of "matter capable of life" (Driesch, 1914). Contrast this view with that of Hypersea theory (see McMenamin and McMenamin, 1994), where living matter and the biospheric living environment are one and the same, cutting out the bio-inert component.

* The domain of phenomena within an organism ("the field of living matter") is different, thermodynamically and chemically, from "the field of the biosphere".

[Editor's note: The manuscript upon which this translation is based carried 28 footnotes by Vernadsky. These are indicated, as here, by an * (or †). All other numbered footnotes are annotations by M. McMenamin (or I.A. Perelman, as noted)].

39 Here again, Vernadsky challenges (without citing) Oparin and Haldane, among others.

40 As put forth by A. I. Oparin (Fox, 1965), "[At] the dawn of European civilization, with the Greek philosophers, there were two clear tendencies in this problem. Those are the Platonic and the Democritian trends, either the view that dead matter was made alive by some spiritual principle or the assumption of a spontaneous generation from that matter, from dead or inert matter.
"The Platonic view has predominated for centuries and, in fact, still continues to exist in the views of vitalists and neo-vitalists."
"The Democritian line was pushed in the background and came into full force only in the seventeenth century in the work of Descartes. Both points of view really differed only in their interpretation of origin, but both of them equally assumed the possibility of spontaneous generation."

izations, by means of which we construct a unified theory of the cosmos. The character of such representations makes them unfruitful when their contents are introduced into the scientific domain.

Mechanistic representations, that on the other hand see merely the simple play of physico-chemical forces in living organisms, are equally fatal to progress in science. They hinder scientific research by limiting its final results; by introducing conjectural constructs based on guesswork,[41] they obscure scientific understanding. Successful conjectures of this sort would rapidly remove all obstacles from the progress of science, but conjectural constructs based on guesswork and their implementation has been linked too closely to abstract philosophical constructs that are foreign to the reality studied by science. These constructs have led to oversimplified analytical approaches, and have thus destroyed the notion of complexity of phenomena.[42] Conjectural constructs based on guesswork have not, thus far, advanced our comprehension of life.

We regard the growing tendency in scientific research to disclaim both these explanations of life, and to study living phenomena by purely empirical processes, as well-founded. This tendency or method acknowledges the impossibility of explaining life, of assigning it a place in our abstract cosmos, the edifice that science has constructed from models and hypotheses.

At the present time, we can approach the phenomena of life successfully only in an empirical fashion, that is, without making unfounded hypotheses. Only in this way can we discover new aspects of living phenomena that will enlarge the known field of physico-chemical forces, or introduce a new principle, axiom, or idea about the structure of our scientific universe. It will be impossible to prove these new principles or notions conclusively, or to deduce them from known axioms, but they will enable us to develop new hypotheses that relate living phenomena to our view of the cosmos, just as understanding of radioactivity connected the view of the cosmos to the world of atoms.

14 The living organism of the biosphere should now be studied empirically, as a particular body that cannot be entirely reduced to known physico-chemical systems. Whether it can be so reduced in the future is not yet clear.[43] It does not seem impossible, but we must not forget another possibility when taking an empirical approach—perhaps this problem, which has been posed by so many learned men of science, is purely illusory.

41 For notes on translation of this passage, see "Editor's Note on Translation and Transliteration."

42 Vernadsky is here challenging simplistic, mechanistic extrapolations in science and in so doing rightly challenges the extensions made of Cartesian- Newtonian mechanics to more complex classes of phenomena. As did Henri Poincaré some decades before, Vernadsky anticipates the problems that chaos theory presents to simple, extrapolation-based mechanistic explanations of phenomena. Vernadsky's intuition is reliable here—recognition of the complexity of the biosphere implies that he had at least an implicit sense of the feedback (cybernetic) dimensions of this field of study, although the language to express these concepts was not developed until shortly after Vernadsky's death. The word cybernetics, from the Greek *kybernetes*, "helmsman," was coined in 1948 by Norbert Wiener.

43 This might seem to make Vernadsky the arch holist (as opposed to reductionist). However, his main point here is that there are probably classes of phenomena that are neither easily nor well explained by inappropriately reductive scientific approaches. Vernadsky's insight on this subject has been decisively vindicated (Mikhailovskii, 1988; Progogine and Stengers, 1988). This makes Vernadsky's scientific approach quite unusual from a Western scientific perspective, for he is a confirmed empiricist who recognizes that holistic approaches will be required to study certain complex entities. His then is not a naïve empiricism, but a sophisticated empiricism in which an empirical approach is utilized to synthesize a scientifically realistic, holistic view of the subject under study. Similar approaches can be identified in the work of the Russian founders of symbiogenesis (Khakhina, 1988; Khakhina, 1992).

Analogous doubts, regarding the governance of all living forms by the laws of physics and chemistry as currently understood, often arise in the field of biology as well.

Even more so than in biology, in the geological sciences we must stay on purely empirical ground, scrupulously avoiding mechanistic and vitalistic constructs. Geochemistry is an especially important case, since living matter and masses of organisms are its principal agents, and it confronts us with living phenomena at every step.

Living matter gives the biosphere an extraordinary character, unique[44] in the universe. Two distinct types of matter, inert[45] and living, though separated by the impassable gulf of their geological history, exert a reciprocal action upon one another. It has never been doubted that these different types of biospheric matter belong to separate categories of phenomena, and cannot be reduced to one. This apparently-permanent difference between living and inert matter can be considered an axiom which may, at some time, be fully established.* Though presently unprovable, this principle must be taken as one of the greatest generalizations of the natural sciences.

The importance of such a generalization, and of most empirical generalizations in science, is often overlooked. The influence of habit and philosophical constructions causes us to mistake them for scientific hypotheses. When dealing with living phenomena, it is particularly important to avoid this deeply-rooted and pernicious habit.

15 There is a great difference between empirical generalizations and scientific hypotheses. They offer quite different degrees of precision. In both cases, we use deductions to reach conclusions, which then are verified by study of real phenomena. In a historical science like geology, verification takes place through scientific observation.

The two cases are different because an empirical generalization is founded on facts collected as part of an inductive research program. *Such a generalization does not go beyond the factual limits, and disregards agreements between the conclusions reached and our representations of nature.* There is no difference, in this respect, between an empirical generalization and a scientifically established fact. Their mutual agreement with our view of nature is not what interests us here, but rather the contradictions between them. Any such contradictions would constitute a *scientific discovery*.

44 So far as we know.

45 Inert matter as used here represents the raw matter, the raw materials of life. Although Vernadsky emphasizes his view that living organisms have never been produced by inert matter, he paradoxically implies that non-living stuff is in some sense alive, or at least has latent life. This should not be confused with any type of mysterious vital force, however; Vernadsky eschewed metaphysical interpretations. He was examining the idea that life has special properties, as old as matter itself, that somehow separated it from ordinary matter (into which it can, by dying, be transformed). Life can expand its realm into inert matter but it was not formed from "nothing."

* The change presently taking place in our ideas regarding mathematical axioms should influence the interpretation of axioms in the natural sciences; the latter have been less thoroughly examined by critical philosophical thought and would constitute a *scientific discovery*.

Certain characteristics of the phenomena studied are of primary importance to empirical generalizations; nevertheless, the influence of all the other characteristics is always felt. An empirical generalization may be a part of science for a long time without being buttressed by any hypothesis. As such, the empirical generalization remains incomprehensible, while still exerting an immense and beneficial effect on our understanding of nature.

But when the moment arrives, and a new light illuminates this generalization, it becomes a domain for the creation of scientific hypotheses, begins to transform our outlines of the universe, and undergoes changes in its turn. Then, one often finds that the empirical generalization did not really contain what was supposed, or perhaps, that its contents were much richer. A striking example is the history of D. J. Mendeleev's great generalization (1869) of the periodic system of chemical elements, which became an extended field for scientific hypothesis after Moseley's discovery[46] in 1915.

16 A hypothesis, or theoretical construction, is fashioned in an entirely different way. A single or small number of the essential properties of a phenomenon are considered, the rest being ignored, and on this basis, a representation of the phenomenon is made. A scientific hypothesis always goes beyond (frequently, far beyond) the facts upon which it is based.[47] To obtain the necessary solidity, it must then form all possible connections with other dominant theoretical constructions of nature, and *it must not contradict them.*[48]

An Empirical Generalization Requires No Verification After It Has been Deduced Precisely from the Facts.

17 The exposition we shall present is based only upon empirical generalizations that are supported by all of the known facts, and not by hypotheses or theories. The following are our beginning principles:

1 During all geological periods (including the present one) there has never been any trace of abiogenesis (direct creation of a living organism from inert matter).

2 Throughout geological time, no azoic (i.e., devoid of life) geological periods have ever been observed.[49]

3 From this follows:

   a) contemporary living matter is connected by a genetic link to the living matter of all former geological epochs; and

46 Actually it was 1913. British physicist H. Moseley studied x-rays emitted by different elements and found that the frequencies in the x-ray spectrum at which the highest intensities occurred varied with the element being studied. In other words, each element has a distinctive x-ray emission 'fingerprint'. This relationship established that the order number of an element in Mendeleev's periodic table (Fersman, 1946) could be established experimentally, and furthermore provided a foolproof method for demonstrating whether or not all the elements of a given region of the table had yet been discovered (Masterton and Slowinski, 1966). These discoveries formed the basis of x-ray energy dispersive (EDS) and wavelength dispersive analytical technology. EDS is frequently used in conjuction with the scanning electron microscope, since the imaging electron beam shot from the tungsten filament in a scanning electron microscope causes the elements in the sample being magnified to radiate their characteristic x-rays. These x-rays are collected by a detector and analysed, thus allowing elemental characterization of specimens being imaged by the scanning electron microscope.

47 It is this extrapolationistic aspect of scientific hypotheses that Vernadsky finds so objectionable.

48 And is far too deductive, in Vernadsky's view, to be the foundation of a reliable scientific methodology. We thus see the profound difference between Western (extrapolations, predictions) and Russian science (assertive scientific generalizations).

49 Again Vernadsky returns to this Huttonesque theme. He really cannot conceive of an azoic Earth. Elsewhere, however, he does admit (Vernadsky, 1939) the possibility that the abiogeneticists could be right (but without ever citing Oparin):

"We cannot shut our eyes, however, to the fact that Pasteur was possibly right, when contemplating in the investigation of these phenomena a way towards the solution of the most important biological problem, and seeking in them the possibility of creation of life on our planet."

Alexei M. Ghilarov (1995) attributes (p. 197) Vernadsky's views on abiogen-

b) the conditions of the terrestrial environment during all this time have favored the existence of living matter, and conditions have always been approximately what they are today.

4   In all geological periods, the chemical influence of living matter on the surrounding environment has not changed significantly; the same processes of superficial weathering have functioned on the Earth's surface during this whole time, and the average chemical compositions of both living matter and the Earth's crust have been approximately the same as they are today.

5   From the unchanging processes of superficial weathering, it follows that the number of atoms bound together by life is unchanged; the global mass of living matter has been almost constant throughout geological time.[50] Indications exist only of slight oscillations about the fixed average.

6   Whichever phenomenon one considers, the energy liberated by organisms is principally (and perhaps entirely) solar radiation. Organisms are the intermediaries in the regulation of the chemistry of the crust by solar energy.

18 These empirical generalizations force us to conclude that many problems facing science, chiefly philosophical ones, do not belong in our investigative domain, since they are not derived from empirical generalizations and require hypotheses for their formulation. For example, consider problems relating to the beginning of life on Earth (if there was a beginning[51]). Among these are cosmogonic models, both of a lifeless era in the Earth's past, and also of abiogenesis during some hypothetical cosmic period.

Such problems are so closely connected with dominant scientific and philosophical viewpoints and cosmogonic hypotheses that their logical necessity usually goes unquestioned. But the history of science indicates that these problems originate outside science, in the realms of religion and philosophy. This becomes obvious when these problems are compared with rigorously established facts and empirical generalizations — the true domain of science. These scientific facts would remain unchanged, even if the problems of biogenesis were resolved by negation, and we were to decide that life had always existed, that no living organism had ever originated from inert matter, and that azoic periods had never existed on Earth. One would be required merely to replace the present cosmogonic hypotheses by new ones, and to apply new scientific and mathematical

esis to his overwhelming empiricism:
"Vernadsky claims that the problem of the origin of life cannot be considered in the framework of empirical science because we know nothing about geological layers that undoubtedly date back to a time when life on the Earth was absent."

In this vein, Vernadsky was fond of citing Redi's Principle of 1669 — *omne vivum e vivo* — "all the living are born from the living" (Vernadsky, 1923, p. 39).

A. Lapo adds here that in 1931 (Lapo, 1980, p. 279) Vernadsky wrote that Redi's principle does not absolutely deny abiogenesis — it only indicates the limits within which abiogenesis does not occur. It is possible that at some time early in Earth's history chemical conditions or states existed on Earth's crust, which are now absent, but which at the time were sufficient for the spontaneous generation of life.

50 This idea of Vernadsky's was controversial even before the 1920's, as pointed out (p. 22) by Yanshin and Yanshina (1988). They note that Vernadsky felt that throughout biological evolution, the forms of living matter had changed but the overall volume and weight of living matter had not changed through time. Convincing proof to the contrary was already available in 1912, when Belgian paleontologist Louis Dollo demonstrated the spread of life from shallow marine waters into oceanic depths and, later, on to land. Vernadsky's error here seems to be a result of the fact that he is completely in the thrall of his slavic variant of substantive uniformitarianism, "the more things change, the more they stay the same." Charles Lyell's western version of extreme substantive uniformitarianism holds that all creatures, including mammals, were present on Earth at a very early time. The Russian version holds biomass as an oscillating constant value through the vastness of geologic time. Dianna McMenamin and I show how the now-recognized increase in biomass over time is a consequence of what Vernadsky elsewhere calls the "pressure of life" (McMenamin and McMenamin 1994). Thus, abandonment of this untenable uniformitarian viewpoint regarding the constancy through geologic time of global biomass does not fundamen-

scrutiny to certain philosophical and religious viewpoints called into question by advances in scientific thought. This has happened before in modern cosmogony.

Living Matter in the Biosphere

19 Life exists only in the biosphere; organisms are found only in the thin outer layer of the Earth's crust, and are always separated from the surrounding inert matter by a clear and firm boundary. Living organisms have never been produced by inert matter. In its life, its death, and its decomposition an organism circulates its atoms through the biosphere over and over again, but living matter is always generated from life itself.

A considerable portion of the atoms in the Earth's surface are united in life, and these are in perpetual motion. Millions of diverse compounds are constantly being created, in a process that has been continuing, essentially unchanged, since the early Archean, four billion years ago.[52]

Because no chemical force on Earth is more constant than living organisms taken in aggregate, none is more powerful in the long run. The more we learn, the more convinced we become that biospheric chemical phenomena never occur independent of life.

All geological history supports this view. The oldest Archean beds furnish indirect indications of the existence of life; ancient Proterozoic rocks, and perhaps even Archean rocks,[53] have preserved actual fossil remains of organisms. Scholars such as C. Schuchert[54] were correct in relating Archean rocks to Paleozoic, Mesozoic, and Cenozoic rocks rich in life. Archean rocks correspond to the oldest-known accessible parts of the crust, and contain evidence that life existed in remotest antiquity at least 1.5 billion years ago.[55] Therefore the sun's energy cannot have changed noticeably since that time; this deduction[56] is confirmed by the convincing astronomical conjectures of Harlow Shapley.[57]

20 It is evident that if life were to cease the great chemical processes connected with it would disappear, both from the biosphere and probably also from the crust. All minerals in the upper crust—the free alumino-silicious acids (clays), the carbonates (limestones and dolomites), the hydrated oxides of iron and aluminum (limonites and bauxites), as well as hundreds of others, are continuously created by the influence of life. In the absence of life, the elements in these minerals would immedi-

tally weaken Vernadsky's other main arguments.

51 This is perhaps the most extreme articulation yet of Vernadsky's substantive uniformitarianism.

52 A. Lapo notes (written communication) that Russian geochemist A. I. Perelman suggested that the following generalization should be called "Vernadsky's Law": "The migration of chemical elements in the biosphere is accomplished either with the direct participation of living matter (biogenic migration) or it proceeds in a medium where the specific geochemical features (oxygen, carbon dioxide, hydrogen sulfide, etc.) are conditioned by living matter, by both that part inhabiting the given system at present and that part that has been acting on the Earth throughout geological history" (Perelman, 1979, p. 215).

53 See Pompeckj, 1928. Indeed as Vernadsky suggests, fossils of microorganisms are now known from Archean rocks.

54 See Schuchert, 1924.

55 Evidence for life is now thought to extend back to 3,800 million years ago; see Mojzsis, Arrhenius, McKeegan, Harrison, Nutman and Friend, 1996; and Hayes, 1996.

56 Now known to be false; the early sun is now thought to have been fainter than today, and yet the planetary surface was paradoxically warmer because of a larger proportion of greenhouse gases (principally carbon dioxide) in the atmosphere.

57 See Shapley, 1927.

ately form new chemical groups corresponding to the new conditions. Their previous mineral forms would disappear permanently, and there would be no energy in the Earth's crust capable of continuous generation of new chemical compounds.[58]

A stable equilibrium, a chemical calm, would be permanently established, troubled from time to time only by the appearance of matter from the depths of the Earth at certain points (e.g., emanations of gas, thermal springs, and volcanic eruptions). But this freshly-appearing matter would, relatively quickly, adopt[59] and maintain the stable molecular forms consistent with the lifeless conditions of the Earth's crust.

Although there are thousands of outlets for matter that arise from the depths of the Earth, they are lost in the immensity of the Earth's surface; and even recurrent processes such as volcanic eruptions are imperceptible, in the infinity of terrestrial time.

After the disappearance of life, changes in terrestrial tectonics would slowly occur on the Earth's surface. The time scale would be quite different from the years and centuries we experience. Change would be perceptible only in the scale of cosmic time, like radioactive alterations of atomic systems.

The incessant forces in the biosphere — the sun's heat and the chemical action of water — would scarcely alter the picture, because the extinction of life would result in the disappearance of free oxygen, and a marked reduction of carbonic acid.[60] The chief agents in the alteration of the surface, which under present conditions are constantly absorbed by the inert matter of the biosphere and replaced in equal quantity by living matter, would therefore disappear.

Water is a powerful chemical agent under the thermodynamic conditions of the biosphere, because life processes cause this "natural" *vadose* water[61] (§89) to be rich in chemically active foci, especially microscopic organisms. Such water is altered by the oxygen and carbonic acid dissolved within it. Without these constituents, it is chemically inert at the prevailing temperatures and pressures of the biosphere. In an inert, gaseous environment, the face of the Earth would become as immobile and chemically passive as that of the moon, or the metallic meteorites and cosmic dust particles that fall upon us.

21 Life is, thus, potently and continuously disturbing the chemical inertia on the surface of our planet. It creates the colors and forms of nature, the associations of animals and plants, and the

58 Here Vernadsky strongly anticipates some of the arguments made later by J. Lovelock, especially the thought that in an abiotic Earth the diatomic nitrogen and oxygen gases will combine to form nitrogen-oxygen compounds (NO_x); Williams, 1997.

59 See Germanov and Melkanovitskaya, 1975.

60 According to A. I. Perelman, the most recent data show that significant amounts of CO_2 are emitted during volcanic eruptions. Evidently, it is no accident that the significance of carbonate deposits abruptly increased after epochs of growing volcanic activity (for example, the Carbonaceous, Jurassic, Paleogene). Note, however, that in the event of the disappearance of life, the atmospheric concentration of CO_2 would rise, while there would be a sharp drop in the percent of carbonate deposits. Indeed, the concentrations of carbon dioxide in the atmospheres of Venus and Mars are very similar (965,000 and 953,000 parts per million volume, respectively), whereas that of Earth is dramatically less (350 parts per million volume; see Williams, 1997, p.110).

61 Vadose water is suspended water in soil or suspended in fragmented rock (regolith), above the level of groundwater saturation. Vernadsky here again demonstrates his marvelous insight, as well as his debt to Dokuchaev (his eacher), as he elucidates the biogeochemical importance of this microbe-rich, high surface-area environment.

creative labor of civilized humanity, and also becomes a part of the diverse chemical processes of the Earth's crust. There is no substantial chemical equilibrium on the crust in which the influence of life is not evident, and in which chemistry does not display life's work.

Life is, therefore, not an external or accidental phenomenon of the Earth's crust. It is closely bound to the structure of the crust, forms part of its mechanism, and fulfills functions of prime importance to the existence of this mechanism. Without life, the crustal mechanism of the Earth would not exist.

22 All living matter can be regarded as a single entity in the mechanism of the biosphere, but only one part of life, *green vegetation*, the carrier of chlorophyll, makes direct use of solar radiation. Through photosynthesis, chlorophyll produces chemical compounds that, following the death of the organism of which they are part, are unstable in the biosphere's thermodynamic field.

The whole living world is connected to this green part of life by a direct and unbreakable link.[62] The matter of animals and plants that do not contain chlorophyll has developed from the chemical compounds produced by green life. One possible exception might be autotrophic bacteria, but even these bacteria are in some way connected to green plants by a genetic link in their past. We can therefore consider this part of living nature as a development that came after the transformation of solar energy into active planetary forces. Animals and fungi accumulate nitrogen-rich substances which, as centers of chemical free energy, become even more powerful agents of change. Their energy is also released through decomposition when, after death, they leave the thermodynamic field in which they were stable, and enter the thermodynamic field of the biosphere.

Living matter as a whole — the totality of living organisms (§160) — is therefore a unique system, which accumulates chemical free energy in the biosphere by the transformation of solar radiation.

23 Studies of the morphology and ecology of green organisms long ago made it clear that these organisms were adapted, from their very beginning, to this cosmic function. The distinguished Austrian botanist I. Wiesner delved into this problem, and remarked, some time ago,[63] that light, even more than heat, exerted a powerful action on the form of green plants . . . "one

62 A partial exception to this general rule was discovered in 1977, the hydrothermal vent biotas of the active volcanic centers of mid-oceanic sea floor spreading ridges (Dover, 1996; Zimmer, 1996). The biotas here are dependent on hydrogen sulfide (normally poisonous to animals) emanating from the volcanic fissures, black smokers and white smokers. Chemosymbiotic bacteria within the tissues of vent biota animals, such as the giant clams and giant tube worms (vestimentiferan pogonophorans), not only detoxify the hydrogen sulfide but utilize it as an energy source in lieu of sunlight. Consider, however, the following from p. 290 of Yanshin and Yanshina (1988):

"Vernadsky considered that the stratified part of the earth's crust (or the lithosphere, as geologists call it) represents a vestige of bygone biospheres, and in that event the granite gneiss stratum was formed as a result of metamorphism and remelting of rocks originating at some point in time under the influence of living matter. Only basalts and other basic magmatic rocks did he regard as deep-seated and not connected genetically with the biosphere." [Although here the connection with the biosphere may simply be a longer period one .–M. McMenamin]

The melting (associated in this case with lithospheric and mantle pressure changes) and eruption of molten rock is probably responsible for exhalation most of the hydrogen sulfide released at mid-ocean ridges. Thus, even with regard to the energy source of the hydrothermal vent biotas (and the incredibly rapid growth of animals living there; see Lutz, 1994,) we may still be considering what is a part of the biosphere *sensu strictu* Vernadsky (E. I. Kolchinsky, 1987; Grinevald, 1996).

63 See Wiesner, 1877.

could say that light molded their shapes as though they were a plastic material."

An empirical generalization of the first magnitude arises at this juncture, and calls attention to opposing viewpoints between which it is, at present, impossible to choose. On the one hand, we try to explain the above phenomenon by invoking internal causes belonging to the living organism, assuming for example that the organism adapts so as to collect all the luminous energy of solar radiation.[64] On the other hand, the explanation is sought outside the organism in solar radiation, in which case the illuminated green organism is treated as an inert mass. In future work the solution should probably be sought in a combination of both approaches. For· the time being the empirical generalization[65] itself is far more important.

The firm connection between solar radiation and the world of verdant creatures is demonstrated by the empirical observation that conditions ensure that this radiation will always encounter a green plant to transform the energy it carries. Normally, the energy of all the sun's rays will be transformed. This transformation of energy can be considered as *a property* of living matter, its *function* in the biosphere. If a green plant is unable to fulfill its proper function, one must find an explanation for this abnormal case.[66]

An essential deduction, drawn from observation, is that this process is absolutely automatic. It recovers from disturbance without the assistance of any agents, other than luminous solar radiation and green plants adapted for this purpose by specific living structures and forms. Such a re-establishment of equilibrium can only be produced in cases of opposing forces of great magnitude. The re-establishment of equilibrium is also linked to the passage of time.

24 Observation of nature gives indications of this mechanism in the biosphere. Let us reflect upon its grandeur and meaning. Land surfaces of the Earth are entirely covered by green vegetation. Desert areas are an exception, but they are lost in the whole.[67] Seen from space, the land of the Earth should appear green, because the green apparatus which traps and transforms radiation is spread over the globe, as continuously as the current of solar light that falls upon it.

Living matter — organisms taken as a whole — is spread over the entire surface of the Earth in a manner analogous to a gas; it produces a specific pressure[68] in the surrounding environment,

[64] In this passage, in which he describes the need to capture light as influencing the morphology of photosynthetic organisms, Vernadsky (following Wiesner) anticipates the research results of both Adolf Seilacher (1985) and Mark McMenamin (1986). The empirical generalization Vernadsky describes here is simply that light influences the shapes of photosynthetic organisms. Either they adapt to maximize light capture, or the light somehow molds the shape of the organisms. The latter suggestion may sound odd but a very similar sentiment was expressed by D'Arcy Wentworth Thompson (1952). In his view, the physical and geometrical contraints of the environment evoke particular shapes from organisms as they evolve, and the array of possible shapes is finite.

[65] That is, Wiesner's inference that light molds plant form.

[66] As for instance in the achlorophyllous Indian Pipe *Monotropa*, which is nourished by linkages to a subterranean network of mycorrhizal mycelia.

[67] In fact, desert areas are clearly identifiable from space.

[68] Here Vernadsky introduces his concept of the "pressure of life." He phrased it succinctly in 1939 (see p. 13) as follows:
"The spreading of life in the biosphere goes on by way of reproduction which exercises a pressure on the surrounding medium and controls the biogenic migration of atoms. It is absent in . . . *inert substance*. The reproduction creates in the biosphere an accumulation of free energy which may be called *biogeochemical energy*. It can be precisely measured."

either avoiding the obstacles on its upward path, or overcoming them. In the course of time, living matter clothes the whole terrestrial globe with a continuous envelope,[69] which is absent only when some external force interferes with its encompassing movement....

This movement is caused by the *multiplication of organisms*, which takes place without interruption,[70] and with a specific intensity related to that of the solar radiation.

In spite of the extreme variability of life, the phenomena of reproduction, growth, and transformation of solar energy into terrestrial chemical energy are subject to fixed mathematical laws. The precision, rhythm, and harmony that are familiar in the movements of celestial bodies can be perceived in these systems of atoms and energy.

The Multiplication of Organisms and Geochemical Energy in Living Matter

25 The diffusion of living matter *by multiplication*, a characteristic of all living matter, is the most important manifestation of life in the biosphere and is the essential feature by which we distinguish life from death. It is a means by which the energy of life unifies the biosphere. It becomes apparent through the *ubiquity of life*, which occupies all free space if no insurmountable obstacles are met. The whole surface of the planet is the domain of life, and if any part should become barren, it would soon be reoccupied by living things. In each geological period (representing only a brief interval in the planet's history), organisms have developed and adapted to conditions which were initially fatal to them. Thus, the limits of life seem to expand with geological time (§119, 122). In any event, during the entirety of geological history life has tended to take possession of, and utilize, all possible space.

This tendency of life is clearly inherent; it is not an indication of an external force, such as is seen, for example, in the dispersal of a heap of sand or a glacier by the force of gravity.

The diffusion of life is a sign of internal energy — of the chemical work life performs — and is analogous to the diffusion of a gas. It is caused, not by gravity, but by the separate energetic movements of its component particles. The diffusion of living matter on the planet's surface is an inevitable movement caused by new organisms, which derive from multiplication and occupy new places in the biosphere; this diffusion is the autonomous energy of life in the biosphere, and becomes known through the

69 See McMenamin and McMenamin, 1994, for examples of this tendency for life to expand its realm.

70 Compare this with the slogan (first pointed out to me by Andrei Lapo) of A. Huxley (1921): "Everything ought to increase and multiply as hard as it can."

transformation of chemical elements and the creation of new matter from them. We shall call this energy *the geochemical energy of life in the biosphere.*

71 Abundant parasites colonizing the tissues of other organisms on land are one of the key characteristics of the land biota.

26 The uninterrupted movement resulting from the multiplication of living organisms is executed with an inexorable and astonishing mathematical regularity, and is the most characteristic and essential trait of the biosphere. It occurs on the land surfaces, penetrates all of the hydrosphere, and can be observed in each level of the troposphere. It even penetrates the interior of living matter, itself, in the form of parasites.[71] Throughout myriads of years, it accomplishes a colossal geochemical labor, and provides a means for both the penetration and distribution of solar energy on our planet.

It thus not only transports matter, but also transmits energy. The transport of matter by multiplication thus becomes a process *sui generis*. It is not an ordinary, mechanical displacement of the Earth's surface matter, independent of the environment in which the movement occurs. The environment resists this movement, causing a friction analogous to that which arises in the motion of matter caused by forces of electrostatic attraction. But movement of life is connected with the environment in a deeper sense, since it can occur only through a gaseous exchange between the moving matter and the medium in which it moves. The more intense the exchange of gases, the more rapid the movement, and when the exchange of gases stops, the movement also stops. This exchange is the *breathing* of organisms; and, as we shall see, it exerts a strong, controlling influence on multiplication. Movement due to multiplication is therefore of great geochemical importance in the mechanisms of the biosphere and, like respiration, is a manifestation of solar radiation.

27 Although this movement is continually taking place around us, we hardly notice it, grasping only the general result that nature offers us — the beauty and diversity of form, color, and movement. We view the fields and forests with their flora and fauna, and the lakes, seas, and soil with their abundance of life, as though the movement did not exist. We see the static result of the dynamic equilibrium of these movements, but only rarely can we observe them directly.

Let us dwell then for a moment on some examples of this movement, the creator of living nature, which plays such an essential yet invisible role. From time to time, we observe the disappearance of higher plant life from locally restricted areas. Forest fires, burning steppes, plowed or abandoned fields, newly-formed islands, solidified lava flows, land covered by volcanic dust or created by glaciers and fluvial basins, and new soil formed by lichens and mosses on rocks are all examples of phenomena that, for a time, create an absence of grass and trees in particular places. But this vacancy does not last; life quickly regains its rights, as green grasses, and then arboreal vegetation, reinhabit the area. The new vegetation enters partially from the outside, through seeds carried by the wind or by mobile organisms; but it also comes from the store of seeds lying latent in the soil, sometimes for centuries.

The development of vegetation in a disturbed environment clearly requires seeds, but even more critical is the geochemical energy of multiplication. The speed at which equilibrium is reestablished is a function of the transmission of geochemical energy of higher green plants.

The careful observer can witness this movement of life, and even sense its pressure,[72] when defending his fields and open spaces against it. In the impact of a forest on the steppe, or in a mass of lichens moving up from the tundra to stifle a forest,[73] we see the actual movement of solar energy being transformed into the chemical energy of our planet.

28 Arthropods (insects, ticks, mites, and spiders) form the principal part of animal living matter on land. In tropical and subtropical regions, the social insects — ants and termites — play the dominant role. The geochemical energy of their multiplication (§37), which occurs in a particular way,[74] is only slightly less than that of the higher green plants themselves.

In termitaries, out of tens and sometimes hundreds of thousands of individuals, only one is endowed with the power of reproduction. This is the queen mother, who lays eggs throughout her life without stopping, and can keep it up for ten years or more. The number of eggs she can lay amounts to millions — some queens have been said to lay sixty eggs per minute with the regularity of a clock ticking seconds.

Multiplication also occurs in swarms, when one part of a generation flies away, with a new queen mother, to a location outside the air space of the founder colony. Instinct serves, with mathematical exactness, for the preservation of eggs instantly carried off by workers, in the flight of swarms, and in the substi-

72 Here Vernadsky injects a qualitative version of his concept of the "pressure of life."

73 Vernadsky makes a veiled reference to Kropotkin (1987) at the beginning of this sentence, and in the next phrase rejects the idyllic connotations of Kropotkin's "mutual aid" theory. Vernadsky's materialist leanings are quite apparent here. Although he never to my knowledge cites it directly, Vernadsky must have been exposed to symbiogenesis theory, for one of his professors was A. S. Famintsyn, founder of Russian plant physiology and one of the chief architects of symbiogenesis theory (Khakhina, 1992). Famintsyn is best known for demostration that photosynthesis can take place under artificial light (Yanshin and Yanshina, 1988; Yanshin and Yanshina, 1989).

74 That is to say, by cooperative breeding (eusociality).

tution of a new queen mother for the old one in case of untimely demise. Marvelously precise laws govern the average values of such quantities as the number of eggs, the frequency of swarms, the numbers of individuals in a swarm, the size and weight of individual insects, and the rate of multiplication of termites on the Earth's surface. These values in turn condition the rate of transmission of geochemical energy by termite motion and expansion. Knowing the numerical constants that define these quantities, we can assign an exact number to the pressure produced on the environment by termites.

This pressure is very high, as is well known by men required to protect their own food supply from termitaries. Had termites met no obstacles in their environment — especially, no opposing forms of life — they would have been able to invade and cover the entire surface of the biosphere in only a few years, an area of over 5×10^8 square kilometers.

75 Actually, 12-15 hours.

29 Bacteria are unique among living things. Although they are the smallest organisms (10^{-4} to 10^{-5} cm in length), they have the greatest rate of reproduction and the greatest power of multiplication. Each divides many times in 24 hours, and the most prolific can divide 63-64 times in a day, with an average interval of only 22-23 minutes between divisions. The regularity of this division resembles that of a female termite laying eggs or a planet revolving around the sun.

Bacteria inhabit a liquid or semi-liquid environment, and are most frequently encountered in the hydrosphere; great quantities also live in soil, and within other organisms. With no environmental obstacles, they would be able to create huge quantities of the complex chemical compounds containing an immense amount of chemical energy, and would be able to do it with inconceivable speed. The energy of this reproduction is so prodigious that bacteria could cover the globe with a thin layer of their bodies in less than 36 hours. Green grass or insects would require several years, or in some cases, hundreds of days.

The oceans contain nearly spherical bacteria, with a volume of one cubic micron. A cubic centimeter could thus contain 10^{12} bacteria. At the rate of multiplication just mentioned, this number could be produced in about 12 hours,[75] starting from a single bacterium. Actually, bacteria always exist as populations rather than as isolated individuals, and would fill a cubic centimeter much more quickly.

The division process takes place at the speed mentioned when conditions are propitious. The bacterial rate of increase drops with temperature, and this drop in rate is precisely predictable.

Bacteria breathe by interaction with gases dissolved in water. A cubic centimeter of water will contain a number of gas molecules much smaller than Loschmidt's number (2.7×10^{19}), and the number of bacteria cannot exceed that of the gas molecules with which they are generatively connected. The multiplication of organized beings is, therefore, limited by respiration and the properties of the gaseous state of matter.

30 This example of bacteria points to another way of expressing the movement in the biosphere caused by multiplication. Imagine the period of the Earth's history when the oceans covered the whole planet. (This is simply a conjecture which was erroneously accepted by geologists). E. Suess[76] dates this "universal sea" or Panthalassa in the Archean Era. It was undoubtedly inhabited by bacteria, of which visible traces have been established in the earliest Paleozoic strata. The character of minerals belonging to Archean beds, and particularly their associations, establish with certainty the presence of bacteria in all the sediments which were lithified to form Archean strata, the oldest strata accessible to geological investigation. If the temperature of the universal sea had been favorable, and there had been no obstacles to multiplication, spherical bacteria (each 10^{-12} cc in volume) would have formed a continuous skin over the Earth's approximately 5.1×10^8 square kilometers in less than thirty six hours.

Extensive films, formed by bacteria, are constantly observed in the biosphere. In the 1890's, Professor M. A. Egounov attempted to demonstrate[77] the existence of a film of sulfurous bacteria, on the boundary of the free oxygen surface[78] (at a depth of about 200 meters), covering an enormous surface area.[79] The research of Professor B. L. Isachenko,[80] performed on N. M. Knipkovitch's 1926 expedition,[81] did not confirm these results; but the phenomenon can nevertheless be observed, at a smaller scale, in other biogeochemically dynamic areas. An example is the junction between fresh and salt water in Lake Miortvoi (Dead Lake)[82] on Kildin Island, where the sediment-water interface is always covered by a continuous layer of purple bacteria.[83]

Other, somewhat larger microscopic organisms, such as plankton, provide a more obvious example of the same kind of phenomenon. Ocean plankton can rapidly create a film cover-

76 See Suess, 1883-1909. This global sea is now called Mirovia.

77 See Egounov, 1897.

78 Also called the "oxygen minimum zone."

79 Based on the depth at which *Thioplaca* mats on the sea floor break up during the Austral winter off the modern coasts of Peru and Chile, storm wave base apparently occurs at 60 meters water depth (see Fossing, et.al., 1995). These mats can indeed be, as per M. A. Egounov's demonstration, of great lateral extent.

80 Boris L. Isachenko was a microbiologist who became heavily involved in the Vernadskian research program. His main interest was the propagation of microorganisms in nature and their role in geological processes, but he also did research in marine microbiology. In 1914 he made the first study of the microflora of the Arctic Ocean as part of a project that was subsequently extended to the Sea of Japan, the Baltic Sea, the Kara Sea, the Sea of Marmora, the Black Sea, the Caspian Sea, and the Sea of Azov. In 1927 he did research on saltwater lakes and medicinal muds. Isachenko also established the role of actinomyces in imparting an earthy odor to water (J. Scamardella, personal communication).

81 Nikolay M. Knipkovitch was a zoologist and ichthyologist. The world's first oceanographic vessel, the *Andrey Pervozvannyy*, was built for his oceanographic expeditions. The voyages of 1922-27 took place in the Sea of Azov and the Black Sea (J. Scamardella, personal communication).

82 Vernadsky was mistaken about the name of this lake: it is Lake Mogilnoe (Grave Lake) (A. Lapo, written communication).

83 See Deriugin, 1925. Vernadsky expressed dismay that the results of K. M. Deriugin's (1878-1936) famous expedition remained only partly published, and urged the Zoological Museum of the Academy of Science to fulfill its scientific and civic duty to fully publish these works (see Vernadsky, 1945, footnote 15).

ing thousands of square kilometers.

84 See Fischer, 1900.

The geochemical energy of these processes can be expressed as the speed of transmission of vital energy to the Earth's surface. This speed is proportional to the intensity of multiplication of the species under consideration. If the species were able to populate the entire surface of the Earth, its geochemical energy would have traversed the greatest possible distance; namely, a great circle of Earth (equal to the length of the equator).

If the bacteria of Fischer[84] were to form a film in Suess's Panthalassic ocean, the speed of transmission of their energy along a great circle would be approximately 33,000 cm/sec., the average speed of movement around the Earth resulting from multiplication starting with one bacterium, for which a complete "tour" of the globe would take slightly less than 36 hours.

The speed of transmission of life, over the maximum distance accessible to it, will be a characteristic constant for each type of homogenous living matter, specific for each species or breed. We shall use this constant to express the geochemical activity of life. It expresses a characteristic both of multiplication, and of the limits imposed by the dimensions and properties of the planet.

31 The speed of transmission of life is an expression not only of the properties of individual organisms, or the living matter of which they are composed, but also of their multiplication as a planetary phenomenon within the biosphere. The size of the planet is an integral part of any such considerations. The concept of *weight* provides an analogy: the weight of an organism on Earth would not be the same as it would be on Jupiter; similarly, the speeds of transmission of life on Earth would be different from the speed observed for the same organism on Jupiter, which has a different diameter.

32 While phenomena of multiplication have been too much neglected by biologists, certain almost unnoticed empirical generalizations about these phenomena have, by their repetition, come to seem obvious. Among these are the following:

1 *The multiplication of all organisms can be expressed in geometrical progressions.* Thus,

$$2^{nD} = N_n$$

where n is the number of days since the start of multiplication; D is the ratio of progression (the number of generations formed in 24 hours, in the case of unicellular organisms multiplying by division); and N_n is the number of individuals formed in n days. D will be characteristic for each homogenous type of living matter (or species). The process is considered infinite: no limits are placed upon n nor N_n in this formula.[85]

2 This potential for infinite growth is nevertheless constrained in the biosphere because *the diffusion of living matter is subject to the law of inertia.*[86] It can be accepted as empirically demonstrated that the process of multiplication is hindered only by external forces. It slows down at low temperatures, and weakens or ceases in the absence of food, of gas to breathe, or of space for the newly born.

In 1858, Darwin[87] and Wallace put this idea in a form familiar to older naturalists, such as C. Linnaeus,[88] G. L. L. Buffon,[89] A. Humboldt,[90] C. G.. Ehrenberg,[91] and K. E. Baer,[92] who had studied the same problem. *If not prevented by some external obstacle, each organism could cover the whole globe and create a posterity equal to the mass of the ocean or the Earth's crust or the planet itself, in a time that is different, but fixed, for each organism.*[93]

3 The specific time required for this is related to the organism's size; small and light organisms multiply more rapidly than large and heavy ones.

33 These three empirical principles portray the phenomenon of multiplication as it never actually occurs in nature, since life is in fact inseparable from the biosphere and its singular conditions. Corrections must be applied to the abstractions for time and space utilized in the above formula.

34 Limitations are imposed upon all quantities that govern the multiplication of organisms, including the maximum number that can be created (N_{max}), the geometrical progression ratio, and the speed of transmission of life. The limits will be determined by the physical properties of the medium in which life exists, and particularly, by the gaseous interchange between organisms and the medium, since organisms must live in a gaseous environment, or in a liquid containing dissolved gases.

35 The dimensions of the planet also impose limitations. The surfaces of small ponds are often covered by floating, green vegetation, commonly duckweed (various species of *Lemna*) in our latitudes. Duckweed may cover the surface in such a closely packed fashion that the leaves of the small plants touch each

85 In other words, the number of individuals of a population after a given number of days is equal to two raised to the power of the growth ratio (the number of generations in a day) times the number of days. The population thus increases rather quickly if the product of the growth ratio and the number of days is large.

86 This is directly analogous to the law of inertia in physics, e. g., a body in motion will remain in motion until acted upon by an external force.

87 Some orthodox practitioners of western-style science have expressed "unease with Darwinism" because it seemed tautological, in other words, difficult to falsify (Ruse, 1988, p. 10). From Vernadsky's point of view, Darwin and Wallace's discovery of natural selection was clearly an extenstion of earlier ideas. But Vernadsky would have been firmly set against the lofty position neo-darwinists have given the role of chance in their evolutionary schema. According (p. 197) to Alexei M. Ghilarov (1995):
"It is understandable, therefore, that despite all his respect for Darwin and Wallace, he considered their concept to be only a general theory of evolution (opposing creationism) rather than a fruitful hypothesis of the origin of species by natural selection. The ideas of stochastic variation, undirectedness, and unpredictability were alien views to Vernadsky" Recall Vernadsky's statement "chance does not exist".

88 See Linnaeus, 1759.

89 See Buffon, 1792.

90 See Humboldt, 1859.

91 See Ehrenberg, 1854.

92 See Baer, 1828, 1876.

93 Here, according to A. I. Perelman, no account is taken of the inner factors, the exhaustion of the capabilities of organisms of a particular species to undergo a final progressive development (compare this with Schindewolf's [translated 1993] concept of senescence), that might

other. Multiplication is hindered by lack of space, and can resume only when empty places are made on the water surface by external disturbances. The maximum number of duckweed plants on the water surface is obviously determined by their size, and once this maximum is reached, multiplication stops. A dynamic equilibrium, not unlike the evaporation of water from its surface, is established. The tension of water vapor and the pressure of life[94] are analogous.

Green algae provide a universally known example of the same process. Algae have a geochemical energy far higher than that of duckweed and, in favorable conditions, can cover the trunks of trees until no gaps are left (§50). Multiplication is arrested, but will resume at the first hint of available space in which to quarter new, individual protococci. The maximum number of individual algae that the surface of a tree can hold is, within a certain margin of error, rigorously fixed.

36 These considerations can be extended to the whole of living nature, although the carrying capacity varies over a wide range. For duckweed or unicellular protococci, it is determined solely by their size; other organisms require much larger surfaces or volumes. In India, the elephant demands up to 30 square kilometers; sheep in Scotland's mountain pastures require about 10,000 square meters; the average beehive needs a minimum of 10 to 15 square kilometers of leafy forest in the Ukraine (about 200 square meters for each bee); 3000 to 15,000 individual plankton typically inhabit a liter of water; 25 to 30 square centimeters is sufficient for ordinary grasses; a few square meters (sometimes up to tens of meters) is needed for individual forest trees.

It is evident that the speed of transmission of life depends on the normal *density* of living matter, an important constant of life in the biosphere.[*] Although this has been little-studied, it clearly applies to continuous layers of organisms, such as duckweed or *Protococcus*, and also applies to a volume completely filled by small bacteria. The concept can be extended to all organisms.

37 With respect to the limitation of multiplication imposed by the dimensions of the planet, *there is evidently a maximum fixed distance over which the transmission of life can take place*; namely, the length of the equator: 40,075,721 meters. If a species were to inhabit the whole of the Earth's surface at its maximum density, it would attain its maximum number of individuals. We

94 Vernadsky's "pressure of life" differs from Lamarck's 1802 concept of the "power of life" (*pouvoir de la vie*; see page 92 in Lamarck, 1964). Lamarck referred to the ability of life to keep living matter in the living state as a "force acting against the tendency of compounds to separate into their constituents" (A. V. Carozzi's footnote 13 in Lamarck, 1964). Vernadsky regards the pressure of life as if he were considering a gas obeying the laws of physics, particularly in its tendency to expand (Wentworth and Ladner, 1972).

Thus Vernadsky's pressurized, expansive properties of life contrast sharply with Lamarck's balancing power of life. Lamarck's view has geological antecedent in the work of Leonardo Da Vinci, who in Folio 36r of *Codex Leicester* described his hypothesis for the relatively constant level of sea water. Da Vinci, following lines of thought begun by Ristoro d'Arezzo, argued that the seas remain at a constant level, and Earth in balance, thanks to subterranean waters that erode Earth's interior, causing caverns to collapse. But for the collapse of caverns, sea water would sink into Earth (Farago, 1996). The collapses prevent the sea from draining completely.

By the seventeenth century the flow of water from cloud to ocean was better understood, leading Sachse de Lowenheimb in his 1664 *Oceanus Macro-microcosmos* to liken hydrospheric circulation to the circulation of blood in the human body.

[*] Vernadsky, 1926b.

95 For example, generations per day.

96 Vernadsky's derivation of biogeochemical constants, from V. I. Vernadsky, 1926c, is as follows:

Δ = optimal number of generations per day

shall call this number (N_{max}) *the stationary number for homogenous living matter*. It corresponds to the maximum possible energy output of homogenous living matter — the maximum geochemical work — and is of great importance for evaluating the geochemical influence of life.

Each organism will reach this limiting number at a speed which is its speed of transmission of life, defined by the formula,

$$V = \frac{13963.3 \, \Delta}{\log N_{max}}$$

If the speed of transmission V remains constant, then obviously the quantity D, which defines the intensity of multiplication[95] (§32), must diminish, as the number of individuals approaches the stationary number and the rate of multiplication slows down.[96]

38 This phenomenon was clearly enunciated 40 years ago by Karl Semper,[97] an accurate observer of living nature, who noted that the multiplication of organisms in small ponds diminished as the number of individuals increased. The stationary number is not actually attained, because the process slows down as the population increases, due to causes that may not be external. The experiments of R. Pearl and his collaborators on *Drosophila* and on fowls (1911-1912) confirm Semper's generalization in other environments.[98]

39 The speed of transmission of life conveys a vivid idea of the geochemical energy of different organisms. As we have seen, it varies widely with the size of the organism, from some 331 meters per second for bacteria (approximately the speed of sound in air), to less than a millimeter (0.9 mm) per second for the Indian elephant. The speeds of transmission of other organisms lie between these two extreme values.

40 In order to determine the energy of life, and the work it produces in the biosphere, both the mass and velocity (or speed of transmission) of the organism must be considered. *The kinetic geochemical energy of living matter is expressed by the formula* $PV^2/2$, where P is the average weight of the organism,* and V is the speed of transmission.

This formula makes it possible to determine the geochemical work that can be performed by a given species, whenever the surface or volume of the biosphere is known.

Attempts to find the geochemical energy of living matter per

k_1 = the greatest dimension (average value) of the organism in cm
V_1 = the velocity of bacteria

For bacteria, take
$\Delta = 64$, $k_1 = 1$ micron = .0001 cm.

Then:

$$V_1 = \frac{13963 \cdot \Delta}{18.71 \, (\log_{10} k_1)}$$

$$V_1 = \frac{13963 \cdot (64)}{18.71 - (-4)}$$

$V_1 = 39,349$ cm/sec = .393 km/sec

(.393 km/sec)(5 sec)(.6214 miles/1 km) = 1.22 miles

Or, in other words, the velocity of bacteria on the surface of the planet works out to be about 1.22 miles in five seconds, assuming of course perfect survivorship of progeny and geometric rates of population increase (conditions which never actually occur in nature). The 13963 multiplier in the numerator of this formula is derived in footnote 19 of Vernadsky, 1989.

The velocity formula used in the above example can be explained as follows. This velocity formula has two forms:

The mean radius of Earth is 6.37 $\times 10^6$ meters, and the surface area is equal to 5.099×10^{14} m^2 or 5.099×10^{18} cm^2. The base ten logarithm of this last number equals 18.707. So, comparing the denominators of the two velocity formulas above,

$18.707 = \log_{10}(k_1) = \log_{10} N_{max}$
$18.707 = \log_{10} N_{max} + \log_{10}(k_1)$
$18.707 = \log_{10}(N_{max})(k_1)$
$10^{18.707} = (k_1)(\log_{10} N_{max})$

$$N_{max} = \frac{10^{18.707}}{k_1}$$

Or, to put it differently, the maximum number of creatures equals their average maximum dimension divided into the surface area of Earth.

97 See for example Semper, 1881.

98 See Pearl, 1912; and Semper, 1881.

* The average weight of a species, P (the average weight of an element of homogenous living matter), logically should be replaced by the average number of atoms in an individual. In the absence of elementary chemical analysis of organisms, this can be calculated only in exceptional cases.

hectare[99] have been made for a long time; for example, in the estimates of *crops*. Facts and theory in this regard are incomplete, but important empirical generalizations have been made. One is that the quantity of organic matter per hectare is both: 1. limited, and; 2. intimately connected with the solar energy assimilated by green plants.

It seems that, in the case of maximum yield, the quantity of organic matter drawn from a hectare of soil is about the same as that produced in a hectare of ocean. The numbers are nearly the same in size, and tend to the same limit, even though soil consists of a layer only a few meters thick, while the life-bearing ocean region is measured in kilometers.[100] The fact that this nearly equal amount of vital energy is created by such different layers can be attributed to the illumination of both surfaces by solar radiation, and probably also to characteristic properties of soil. As we shall see, organisms that accumulate in the soil (microbes) possess such an immense geochemical energy (§155) that this thin soil layer has a geochemical effect comparable to that of the ocean, where the concentrations of life are diluted in a deep volume of water.

41 The kinetic geochemical energy $PV^2/2$, concentrated per hectare, may be expressed by the following formula:[101]

$$A_1 = \left(\frac{PV^2}{2}\right) \times \left(\frac{10^8}{K}\right) = \frac{(PV^2)\,(N_{max})}{2(5.10065 \times 10^{18})}$$

where $10^8/K$ is the maximum number of organisms per hectare (§37); K is the coefficient of density of life (§36); N_{max} is the stationary number for homogenous living matter (§37); and 5.10065×10^{18} is the area of the Earth in square centimeters. Characteristically, this quantity seems to be a constant for protozoa, for which the formula gives $A_1 = (PV^2/2) \times (10^8/K) = a \times (3.51 \times 10^{12})$ in CGS units. The coefficient a is approximately one.[*]

This formula shows that the kinetic geochemical energy is determined by the velocity V, and is thus related to the organism's weight, size, and intensity of multiplication. In relation to Δ, V can be expressed as

$$V = \frac{(46,383.93)\,(\log 2)\,(\Delta)}{18.70762 - \log K} \quad \text{[in CGS units]}^{\dagger},$$

in which the constants are related to the size of the Earth. The largest known value for V is 331 meters per second; and for Δ, about 63 divisions per day.[102]

This formula shows that the size of the planet, alone, cannot

99 A large quantity of corroborative data for natural vegetation is found in the book by Rodin and Basilevich, 1965.

100 Although the notion was important to Vernadsky (possibly because it demonstrated that the transformative power of life was as potent on land as in the sea), this assertion that land and sea biomass are roughly equal is not valid. Upward transport by vascular plants of fluid and nutrients allows the land biota to far outstrip the marine biota (by approximately two orders of magnitude) in terms of overall biomass. (McMenamin and McMenamin, 1993; McMenamin and McMenamin, 1994). Annual productivity on a per square meter basis is about four times greater on the land than in the sea.

101 This formula calculates the value A_1, the geochemical energy of a particular species of organism concentrated on a given patch of Earth's surface area. It is calculated by dividing the product of the geochemical energy of that species $(PV^2/2)$ and its maximum abundance on Earth (N_{max}) by the surface area of Earth. The "2" in the denominator of the final quotient is from the denominator of $PV^2/2$. The calculation is an interesting and unusual way to describe the bioenergetics of organisms.

* Corresponding to the density of protozoan protoplasm, which, by recent measurements (see Leontiev, 1927), is about 1.05. The quicker the multiplication, the more intense the respiration.

† This expression V applies for all organisms, and not just for protozoans. For all other groups, such as higher animals and plants, the expression A_1 has another, *lesser* value, as a result of profound differences between the metabolism and organization of complex creatures (such as animals and plants) and unicellular protists. I cannot here delve into examination of these complex and important distinctions.

[Editor's note: This footnote appears in the 1989 edition but is cryptic because Vernadsky makes just such a comparison in sections to follow. Perhaps he meant that he did not account for the actual limits imposed upon V and Δ. Can these quantities attain higher values, or does the biosphere impose limits upon them?[103] An obstacle that imposes maximum values upon these constants does, in fact, exist; namely, the gaseous exchange that is essential for the life and multiplication of organisms.

42 Organisms cannot exist without exchange of gases — *respiration* — and the intensity of life can be judged by the rate of gaseous exchange.

On a global scale, we must look at the general result of respiration, rather than at the breathing of a single organism. The respiration of all living organisms must be recognized as part of the mechanism of the biosphere. There are some long-standing empirical generalizations in this area, which have not yet been sufficiently considered by scientists.

The first of these is that *the gases of the biosphere are identical to those created by the gaseous exchange of living organisms*. Only the following gases are found in noticeable quantities in the biosphere, namely oxygen, nitrogen, carbon dioxide, water, hydrogen, methane, and ammonia. This cannot be an accident. The free oxygen in the biosphere is created solely by *gaseous exchange in green plants*,[104] and is the principal source of the free chemical energy of the biosphere. Finally, *the quantity of free oxygen* in the biosphere, equal to 1.5×10^{21} grams (about 143 million tons[105]) is of the same order as the existing quantity of living matter,[106] independently estimated at 10^{20} to 10^{21} grams.[107] Such a close correspondence between terrestrial gases and life strongly suggests[108] that the breathing of organisms has primary importance in the gaseous system of the biosphere; in other words, *it must be a planetary phenomenon*.

43 The intensity of multiplication, and likewise the values of V and Δ, cannot exceed limits imposed by properties of gases, because they are determined by gaseous exchange. We have already shown (§29) that the number of organisms that can live in a cubic centimeter of any medium must be less than the number of molecules of gas it contains (Loschmidt's number; 2.716×10^{19} at standard temperature and pressure*). If the velocity V were greater than 331 meters per second, the number of organisms smaller than bacteria (i.e., with dimensions 10^{-5} centimeters or smaller) would exceed 10^{19} per cubic centimeter. Due to respiration, the number of organisms that exchange gas mole-

102 This formula is derived in footnote 22 of Vernadsky, 1989. It is in a sense redundant; Vernadsky includes it as a demonstration, to confirm for readers that the speed of transmission of life (V) may be expressed as a function of the generations per day (Δ), the size of the organisms in question (K), *and* the dimensions of Earth.

103 Alexei M. Ghilarov (1995) had this (p. 200) to say about Vernadsky's calculation:
"Vernadsky claimed that the rate of natural increase and dispersal of any organism must be related to the area of the Earth's surface, to the length of the equator, to the duration of one rotation of the Earth on its axis, and other *planetary characteristics*. . . . Emphasizing that "all organisms live on the Earth in restricted space which is *of the same size* for all of them" Vernadsky . . . . simply implies that all organisms inhabit a common planet of a finite size [italics his]."
But Ghilarov misses the main point of Vernadsky's mathematical demonstration. For Vernadsky, the size of Earth is invariant. The main variables, which are constant for any species, are Δ (often expressed in generations per day) and K (the organism's size). So the only thing that truly varies, and thus determines the geochemical energy and the velocity or speed of transmission of life, is an organism's respiratory rate (the rate of exchange of gases in air or as dissolved gases in water). For Vernadsky, respiration is the key to understanding any species of organisms, for respiration is the fundamental process linking the organism to the rest of the biosphere. An organism's respiring surfaces represent the interface across which living matter and bio-inert matter interact.

104 In 1856 C. Koene [citation unknown] hypothesized that atmospheric oxygen was the result of photosynthesis. Vernadsky gave this idea special attention, and from the perspective of geochemistry (Voitkevich, Miroshnikov, Povarennykh, Prokhorov, 1970). The Keene hypothesis was accepted without much com-

cules would have to increase as their individual dimensions decreased. As their dimensions approached that of molecules, the speed would rise to improbable values and become physically absurd.

Breathing clearly controls the whole process of multiplication on the Earth's surface. It establishes mutual connections between the numbers of organisms of differing fecundity, and determines, in a manner analogous to temperature, the value of Δ that an organism of given dimensions can attain. Limitations to the ability to respire are the primary impediment to the attainment of maximum population density.

Within the biosphere, there is a desperate struggle among biospheric organisms, not only for food, but also for air; and the struggle for the latter is the more essential, for it controls multiplication. Thus respiration (or breathing) controls maximal possible geochemical energy transfer per hectare surface area.

44 On the scale of the biosphere, the effect of gaseous exchange and the multiplication it controls is immense. Inert matter exhibits nothing even remotely analogous, since any living matter can produce an unlimited quantity of new living matter.

The weight of the biosphere is not known, but it is certainly only a tiny fraction of the total weight of the Earth's crust (or even of the 16-20 kilometers that participate in geochemical cycles accessible to direct study) (§78). The weight of the top 16 kilometers is 2×10^{25} grams, but if there were no environmental obstacles, a much larger amount of living matter could be created by multiplication in a negligible span of geological time. The cholera vibrio and the bacterium *E. coli* could yield the above mass in 1.6 to 1.75 days. The green diatom *Nitzchia putrida*, a mixotrophic organism of marine slimes which consumes decomposed organic matter and also uses solar radiation in its chloroplasts, could produce 2×10^{25} grams in 24.5 days. (This is one of the fastest growing organisms, possibly because it utilizes already existing organic matter.)

The Indian elephant, having one of the slowest multiplication rates, could produce the same quantity of matter in 1300 years, a short moment in the scale of geological time. Further along the growth curve, of course, the elephant could produce the same mass in days.[109]

45 Obviously, no organism produces such quantities of matter in the real world. There is nothing fantastic, however, about dis-

plaint since it was known that plants release oxygen (see Van Hise, 1904, p. 949; "it is suspected that a considerable percentage of the oxygen now in the atmosphere could be thus be accounted for" [i.e., by photosynthesis]), but Vernadsky was the first to demonstrate the biogenic origin of atmospheric oxygen in its global entirety (Vernadsky, 1935; see also Oparin, 1957, p. 157). For a discussion of the current status of the problem see Molchanov and Pazaev, 1996. [The citation for Koene, 1856, and few others noted in the text elsewhere, have not been located. If anyone reading this text is familiar with this or other unknown or incomplete citations noted, please provide the information to the publisher, Peter N. Nevraumont, Nevraumont Publishing Company, 16 East 23rd Street, New York, New York 10010, and it will be included in future editions.]

105 The current estimate for the mass of the atmosphere is 5×10^{24} grams. Multiplying this value by the weight percent of oxygen in the atmosphere (22.87% of atmospheric mass assuming 2% by volume water vapor in air; see Gross, 1982; and Levine, 1985) gives an atmospheric oxygen content of 1.143×10^{24} grams. Vernadsky's value is too low by at least three orders of magnitude. He must have badly underestimated the mass of the atmosphere.

106 It is difficult to verify or reject Vernadsky's assertion here. The total biomass of Earth is still poorly known, as a result of uncertainties as to the total biomass of subterranean bacteria.

107 According to recent data, the total biomass of Earth averages 7.5 $\times 10^{17}$ grams of organic carbon (Romankevich, 1988). Vernadsky apparently meant total biomass, whereas Romankevich's data include only organic carbon.

108 Because of the vast amount of free oxygen in the atmosphere.

*Microbes live in a gaseous environment having this number of molecules at 0° and 760 mm pressure. In the presence of bacteria, the number of gaseous molecules must be less. A cubic centimeter of liquid containing

placements of mass of this order resulting from multiplication in the biosphere. Exceptionally large masses of organisms are actually observed in nature. There is no doubt that life creates matter at a rate several times greater than 10^{25} grams per year.[110] The biosphere's 10^{20} to 10^{21} grams of living matter is incessantly moving, decomposing, and reforming. The chief factor in this process is not growth, but multiplication. New generations, born at intervals ranging from tens of minutes to hundreds of years, renew the substances that have been incorporated into life.

Because enormous amounts of living matter are created and decomposed every 24 hours, the quantity which exists at any moment is but an insignificant fraction of the total created in a year.

It is hard for the mind to grasp the colossal amounts of living matter that are created, and that decompose, each day, in a vast dynamic equilibrium of death, birth, metabolism, and growth. Who can calculate the number of individuals continually being born and dying? It is more difficult than Archimedes' problem of counting grains of sand — how can they be counted when their number varies and grows with time? The number that exists, in a time brief by human standards, certainly exceeds the grains of sand in the sea by a factor of more than 10^{25}.

Photosynthetic Living Matter

46 The amount of living matter in the biosphere (10^{20} to 10^{21} grams) does not seem excessively large, when its power of multiplication and geochemical energy are considered.

All this matter is generatively connected with the living green organisms that capture the sun's energy. The current state of knowledge does not allow us to calculate the fraction of all living matter that consists of green plants, but estimates can be made. While it is not certain that green living matter predominates on the Earth as a whole, it does seem to do so on land.[111] It is generally accepted that animal life predominates (in volume) in the ocean. But even if heterotrophic animal life should be found to be the greater part of all living matter, its predominance cannot be large.

Are the two parts of living matter — photoautotrophic and heterotrophic — nearly equal in weight? This question cannot now be answered,[112] but it can be said that estimates of the weight of green matter, alone, give values of 10^{20} to 10^{21} grams, which are the same in order of magnitude as estimates for living matter *in toto*.

microbes must contain fewer than 10^{19} molecules; it cannot at the same time contain a like number of microbes.

109 An interesting comparison (suggested by Peter N. Nevraumont) may be made between Vernadsky's and Darwin's interpretation of the rate of increase of elephants, the slowest breeding animals. Whereas Vernadsky emphasized the biogeological accumulation of a quantity of elephant "matter," Darwin emphasized the geometrical rate of increase in the number of individual elephants in the struggle for existence, calculating that within 740-750 years a single breeding female could theoretically produce nineteen million offspring (Darwin, 1963, p.51).

110 Recent calculations show the total biomass production of Earth averages 1.2×10^{17} grams of organic carbon per year (Romankevich, 1988; Schlesinger 1991). In energetic terms, solar energy is fixed in plants by photsynthesis at a net rate of about 133 TW (10^{12} W; see Lovins, Lovins, Krause, and Bach, 1981).

111 On land, the total biomass of autotrophs is nearly a hundred times as large as the biomass of heterotrophs (738×10^{15} versus 8.10×10^{15}, respectively). See Romankevich, 1988.

112 The present answer to this question would be "no." Total photoautotrophic biomass on Earth is 740×10^{15} grams of organic carbon, whereas total heterotrophic biomass is only about 10×10^{15} grams of organic carbon. See Romankevich, 1988; and Schlesinger 1991.

47 Solar energy transformers on land are structured quite differently from those in the sea. On dry land, phanerogamous,[113] herbaceous plants predominate. Trees probably represent the greatest fraction, by weight, of this vegetation; green algae and other cryptogamous plants (principally protista) represent the smallest fraction. In the ocean, microscopic, unicellular green organisms predominate; grasses like *Zostera* and large algae constitute a smaller portion of green vegetation, and are concentrated along shores in shallow areas accessible to sunlight. Floating masses of them, like those in the Sargasso Sea, are lost in the immensity of the oceans.

Green metaphytes[114] predominate on land; in this group, the grasses multiply at the greatest speed and possess the greatest geochemical energy, whereas trees appear to have a lower velocity. In the ocean, green protista have the highest velocity.

The speed of transmission v, for metaphytes, probably does not exceed a few centimeters per second. Green protista have a speed of thousands of centimeters per second, besting the metaphytes by hundreds of times with regard to power of multiplication, and clearly demonstrating the difference between marine and terrestrial life. Although green life is perhaps less dominant in the sea than on the soil, the total mass of green life in the ocean exceeds that on land because of the larger size of the ocean itself. The green protista of the ocean are the major agents in the transformation of luminous solar energy into chemical energy on our planet.[115]

48 The energetic character of green vegetation can be expressed quantitatively in a way that shows the distinction between green life on land and in the sea. The formula $Nn = 2^{n\Delta}$ gives the growth (α) of an organism in 24 hours due to multiplication. If we start with a single organism ($n = 1$ on the first day), we shall have:

$$2^{\Delta} - 1 = \alpha$$
$$2^{\Delta} = \alpha + 1 \text{ and } 2^{n\Delta} = (\alpha + 1)^n$$

The quantity a is a constant for each species; it is the number of individuals that will grow in 24 hours starting from a single organism. The magnitude $(\alpha + 1)^n$ is the number of individuals created by multiplication on the nth day: $(\alpha + 1)^n = Nn$.

The following example shows the significance of these numbers. The average multiplication of plankton, according to Lohmann, can be expressed by the constant $(\alpha + 1) = 1.2996$, taking into account the destruction and assimilation of the

113 That is, those with visible reproductive organs such as flowers and cones.

114 Land plants, members of kingdom Plantae.

115 Vernadsky's assertions here have not been borne out. The mass of photoautotrophs on land (738 × 10^{15} grams of organic carbon) vastly outweighs the mass of photoautotrophs in the sea (1.7 × 10^{15} grams of organic carbon) (Romankevich, 1988). This discrepancy has recently been attributed to upward nutrient transport by vascular plants on land (McMenamin and McMenamin, 1994).

plankton by other organisms. The same constant for an average crop of wheat in France is 1.0130. These numbers correspond to the ideal average values for wheat or plankton after 24 hours of multiplication. So the ratio of the number of plankton individuals to those of wheat is

$$\frac{1.2996}{1.0130} = 1.2829 = \delta$$

This ratio is multiplied every 24 hours by δ, being δ^n after n days.

On the 20th day, the value would be 145.8; on the hundredth, the number of plankton would exceed that of wheat plants by a factor of 6.28×10. After a year, neglecting the fact that the multiplication of wheat is arrested for several months, the ratio of the populations (δ^{365}) attains the astronomical figure of 3.1×10^{39}. The initial difference between the full-grown herbaceous plant (weighing tens of grams) and the microscopic plankton (weighing 10^{-10} to 10^{-6} grams) is dwarfed by the difference in intensity of multiplication.

The green world of the ocean gives a similar result, due to the speed of circulation of its matter.[116] The force of solar radiation allows it to create a mass equivalent in weight to the Earth's crust (§44) in 70 days or less. Herbaceous vegetation on land would require years to produce this quantity of matter — in the case of *Solanum nigrum*, for example, five years.

These figures, of course, do not give a correct perspective of the relative roles of herbaceous vegetation and green plankton in the biosphere, because in this method of comparison the difference grows enormously with time. In the five-year span mentioned above for *Solanum nigrum*, for example, the amount of green plankton that could be produced would be hard to express in conceivable figures.

49 It is not accidental that living green matter on land differs from that in the sea, because the action of solar radiation in a transparent, liquid medium is not the same as on solid, opaque Earth. The world of plankton controls geochemical effects in the oceans, and also on land wherever aqueous life exists.

The difference in energy possessed by these two kinds of living matter is represented by the quantity δ^n, and also by the mass (m) of the individuals created. This mass is determined by the product of the number of individuals created, and their average weight (P): $m = P(1 + \alpha)^n$. Small organisms would have the advantage over large ones, energetically-speaking, only if they really could produce a larger mass in the biosphere.

116 The point, then, of the immediately preceding mathematical calculations is that differences in the intensity of multiplication between complex, larger organisms and smaller ones (differences which grow astronomically large with passage of time) are a direct result of the differing relative respiratory surfaces of large and small organisms, respectively. Smaller organisms have a much greater surface to volume ratio (and hence greater respiratory surface area to volume ratio) in comparison to larger, more complex organisms. This accounts for, all other things being equal, the disparity between large and small organisms in their velocity or speed of transmission (V), and then of course the much greater disparity (because it is calculated using the square of V) between their respective geochemical energies ($PV^2/2$). In comparisons of this sort, the microbes outperform more complex organisms by an overwhelming margin, assuming they are able to create sufficient biomass. Vernadsky thus provides the quantification required for comparisons of geochemical energy between species.

Any system reaches a stable equilibrium when its free energy is reduced to a minimum under the given conditions; that is, when all work possible in these conditions is being produced. All processes, of both the biosphere and the crust, are determined by conditions of equilibrium in the mechanical system of which they are a part.[117]

Solar radiation and the living green matter of the biosphere, taken together, constitute a system of this kind. When solar radiation has produced the maximum work, and created the greatest possible mass of green organisms, this system has reached a stable equilibrium.

Since solar radiation cannot penetrate deeply into solid earth, the layer of green matter it creates there is limited in thickness.

The environment gives all the advantages to large plants, grasses and trees as compared to green protista. The former create a larger quantity of living matter, although they take a longer time to do it. Unicellular organisms can produce only a very thin layer of living matter on the land surface, and soon reach a stationary state (§37) at the limits of their development. In the system of solar radiation and solid earth taken as a whole, unicellular organisms are an unstable form,[118] because herbaceous and wooded vegetation, in spite of their smaller reserve of geochemical energy, can produce much more work, and a greater quantity of living matter.

50 The effects of this are seen everywhere. In early spring, when life awakens, the steppe becomes covered in a few days by a thin layer of unicellular algae (chiefly larger cyanobacteria and algae such as *Nostoc*). This green coating develops rapidly, but soon disappears, making room for the slower-growing herbaceous plants. Due to the properties of the opaque earth, the grass takes the upper hand, although the *Nostoc* has more geochemical energy. Everywhere, tree bark, stones, and soil are rapidly covered by fast-developing *Protococcus*. In damp weather, these change in only a few hours from cells weighing millionths of a milligram into living masses weighing decigrams or grams. But even in the most favorable conditions, their development soon stops. As in Holland's sycamore groves, tree trunks are covered by a continuous layer of *Protococcus* in stable equilibrium, further development of which is arrested by the opacity of the matter on which they live. The fate of their aqueous cousins, freely developing in a transparent medium hundreds of meters deep, is quite different.

117 Vernadsky offers here a plausible explanation for the stability of Earth's climate; it represents a minimum energy configuration. See McMenamin, 1997b.

118 Vernadsky is saying here that unicellular organisms reach a stable state only in the absence of competing large plants, as might have been the case during the Precambrian. Large plants today prevent the microbes from reaching this stable or stationary state.

Trees and grasses, growing in a new transparent medium (the troposphere), have developed forms according to the principles of energetics and mechanics. Unicellular organisms may not follow them on this path. Even the appearance of trees and grasses, the infinite variety of their forms, displays the tendency to produce maximum work and to attain maximum bulk of living matter.

To reach this aim, they created a new medium for life — the atmosphere.[119]

51 In the ocean, where solar radiation penetrates to a depth of hundreds of meters, unicellular algae, with higher geochemical energy, can create living matter at an incomparably faster rate than can the plants and trees of land. In the ocean, solar radiation is utilized to its utmost. The lowest grade of photosynthetic organism has a stable vital form: this leads to an exceptional abundance of animal life, which rapidly assimilates the phytoplankton, enabling the latter to transform an even greater quantity of solar energy into living mass.

52 Thus, solar radiation as the carrier of cosmic energy not only initiates its own transformation into terrestrial chemical energy, but also actually creates the transformers themselves. Taken together, these make up living nature, which assumes different aspects on land and in the water.

The establishment of the life forms is thus in accordance with the way solar (cosmic) energy changes the structure of living nature, by controlling the ratio of autotrophic to heterotrophic organisms. A precise understanding of the laws of equilibrium that govern this is only now beginning to appear.

Cosmic energy determines the pressure of life (§27), which can be regarded as the transmission of solar energy to the Earth's surface.[120] This pressure arises from multiplication, and continually makes itself felt in civilized life. When man removes green vegetation from a region of the Earth, he changes the appearance of virgin nature, and must resist the pressure of life, expending energy and performing work equivalent to this pressure. If he stops this defense against green vegetation, his works are swallowed up at once by a mass of organisms that will repossess, whenever and wherever possible, any surface man has taken from them.

This pressure is apparent in the *ubiquity of life*. There are no regions which have always been devoid of life. We encounter

119 As Vernadsky later notes, Dumas and Boussingault in 1844 considered life to be an "appendage of the atmosphere." More recently it has been argued that the atmosphere is a nonliving product of life, like a spider's web, a view Vernadsky seems to have anticipated here.

120 This is a remarkable passage. Thus life not only plays the role of horizontal transmission of matter, but also horizontal transmission of vertically-incoming solar energy.

vestiges of life on[121] the most arid rocks, in fields of snow and ice, in stony and sandy areas. Photosynthetic organisms are carried to such places mechanically; microscopic life is constantly born, only to disappear again; animals pass by, and some remain to live there. Some richly-animated concentrations of life are observed, but not as a green world of transformers. Birds, beasts, insects, spiders, bacteria, and sometimes green protista make up the populations of these apparently inanimate regions, which are really *azoic* only in comparison with the "immobile" green world of plants. These regions can be likened to those of our latitudes where green life disappears, temporarily, beneath a clothing of snow, during the winter suspension of photosynthesis.

Phenomena of this sort have existed on our planet throughout geological history, but always to a relatively limited extent. Life has always tended to become master of apparently lifeless regions, adapting itself to ambient conditions. Every empty space in living nature, no matter how constituted, must be filled in the course of time. Thus, new species and subspecies of flora and fauna will populate azoic areas, newly formed land areas, and aquatic basins. It is curious and important to note that the structures of these new organisms, as well as the structures of their ancestors, contain certain preformed properties that are required for the specific conditions of the new environment.[122] This morphological preformation and the ubiquity of life are both manifestations of the energetic principles of the pressure of life.

Azoic surfaces, or surfaces poor in life, are limited in extent at any given moment of the planet's existence. But they always exist, and are more evident on land than in the hydrosphere. We do not know the reason for the restrictions they impose on vital geochemical energy; nor do we know whether there exists a definite and inviolable relationship between the forces on the Earth that are opposed to life, on one hand, and the life-enhancing and not yet fully understood force of solar radiation on the other.

53 The ways in which green vegetation has adapted so as to attract cosmic energy can be seen in many ways. Photosynthesis takes place principally in tiny plastids, which are smaller than the cells they occupy. Myriads of these are dispersed in plants, to which they impart the green color.

Examination of any green organism will show how it is both generally and specifically adapted to attract *all* the luminous

[121] Or in the rocks, as in the case of cryptoendolithic Antarctic lichens.

[122] Here Vernadsky alludes to what is today called preadaptation or exaptation. Evolutionists now feel that organisms have no ability to anticipate environmental change. But once such change has occurred (or sometimes without any such change), organs useful for one function can switch their function and provide the organism bearing them with a new adaptation—hence the appearance that they were "preadapted." See Cuénot, 1894a, 1894b and 1925.

radiation accessible to it. Leaf size and distribution in plants is so organized that not a single ray of light escapes the microscopic apparatus which transforms the captured energy. Radiation reaching Earth is gathered by organisms lying in wait. Each photosynthetic organism is part of a mobile mechanism more perfect than any created by our will and intelligence.

The structure of vegetation attests to this. The surface of leaves in forests and prairies is tens of times larger than the area of the ground they cover. The leaves in meadows in our latitudes are 22 to 38 times larger in area; those of a field of white lucerne are 85.5 times larger; of a beech forest, 7.5 times; and so on, even without considering the organic world that fills empty spaces rapidly with large-sized plants. In Russian forests, the trees are reinforced by herbaceous vegetation in the soil, by mosses and lichens which climb their trunks, and by green algae which cover them even under unfavorable conditions.[123] Only by great effort and energy can man achieve any degree of homogeneity in the cultivated areas of the Earth, where green weeds are constantly shooting up.

This structure was strikingly demonstrated in virgin nature before the appearance of man, and we can still study its traces. In the uncultivated regions of "virgin steppe" which survive in central Russia, one can observe a natural equilibrium that has existed for centuries, and could be reestablished everywhere if man did not oppose it. J. Paczoski[124] has described the steppe of "Kovyl" or needle grass (*Stipa capillata*) of Kherson: "It gave the impression of a sea; one could see no vegetation except the needle grass[125] which rose as high as a man's waist and higher. The mass of this vegetation covered the land almost continuously, protecting it by shade and helping it to conserve the humidity of the soil, so that lichens and mosses were able to grow between the tufts of the leaves and remain green at the height of summer."[126]

Earlier naturalists have similarly described the virgin savannas of Central America. F. d'Azara (1781-1801) writes[127] that the plants were "so thick that the earth could only be seen on the roads, in streams, or in gulleys."

These virgin steppes and savannas are exceptional areas that have escaped the hand of civilized man, whose green fields have largely replaced them.

In our latitudes, vegetation lives with a periodicity controlled by an astronomical phenomenon—the rotation of the Earth around the sun.

123 The wording at the end of this sentence suggests, as inferred earlier, exposure to the ideas of the symbiogeneticists and to the ideas that organisms of different species can aid one another.

124 See Paczoski, 1908.

125 Needle grass, or *Stipa capillata*, is referred to in Russian as *tyrsa*. It is a tough grass with sharp leaves. Most species of the genus *Stipa* are perennials, and they are a characteristic plant of the steppes throughout the world.

126 This same observation strongly influenced Kropotkin.

127 See d'Azara, 1905.

54 The same picture of saturation of the Earth's surface by green matter can be observed in all other phenomena of plant life: forest stands of tropical and subtropical regions, the taiga of septentrional and temperate latitudes, savannas and tundras. These are the coating with which green matter permanently or periodically covers our planet, if the hand of man has not been present. Man, alone, violates the established order; and it is a question whether he diminishes geochemical energy, or simply distributes the green transformers in a different way.[128]

Grouped vegetation and isolated plants of many forms are so arranged as to capture solar radiation, and to prevent its escaping the green-chlorophyll plastids.

Generally radiation cannot reach any locality of the Earth's surface[129] without passing through a layer of living matter that has multiplied by *over one hundred times* the surface area that would otherwise be present if life were absent from the site.

55 Land comprises 29.2 percent of the surface of the globe; all the rest is occupied by the sea, where the principal mass of green living matter exists and most of the luminous solar energy is transformed into active chemical energy.

The green color of living matter in the sea is not usually noticed, since it is dispersed in myriad microscopic, unicellular algae. They swim freely, sometimes in crowds, and at other times spread out over millions of square miles of ocean. They can be found wherever solar radiation penetrates, up to 400 meters water depth, but mostly between 20 and 50 meters from the surface, rising and sinking in perpetual movement. Their multiplication varies according to temperature and other conditions, including the rotation of the planet around the sun.

Incident sunlight is undoubtedly utilized in full by these organisms. Green algae, cyanobacteria, brown algae, and red algae succeed each other in depth in a regular order.[130] The red phycochromaceae use the blue rays, the final traces of solar light not absorbed by water. As W. Engelmann has shown,[131] all these algae, each type with its own particular color, are adapted to produce maximum photosynthesis in the luminous conditions peculiar to their aqueous medium.[132]

This succession of organisms with increasing depth is a ubiquitous feature of the hydrosphere. In shallows, or in special structures linked with geological history such as the Sargasso Sea, the plankton, though invisible to the naked eye, are intensified by immense, floating fields or forests of algae and plants,

128 Vernadsky never makes up his mind about the true role of humans in the biosphere (Vernadsky, 1945).

129 Here Vernadsky means any forested locality of Earth's surface.

130 The farbstreifensandwatt, or "color-striped sand," is a within-sediment bacterial and algal community showing this same type of stratified succession. The microbial photosynthesizers of the farstreifensandwatt partition the light by wavelength and intensity as it passes through the sand layer. See Schulz, 1937; Hoffmann, 1949; and Krumbien, Paterson and Stal, 1994.

131 See Engelmann 1984; and 1861.

132 They also have distinctive characteristics of metal accumulation (Tropin and Zolotukhina, 1994).

some of them very large. These are chemical laboratories, with energy more powerful than the most massive forests of solid earth. The total surface they occupy, however, is relatively small — only a few percent of the surface area occupied by the plankton.

56 Thus, we see that the hydrosphere, a majority of the planetary surface, is always suffused with an unbroken layer of green energy transformers, as is most of the continental area in the appropriate seasons. Places poor in life, such as glaciers, and azoic regions constitute only 5 to 6 percent of the total surface area; with this taken into account, the layer of green matter still has a surface area far greater than that of the Earth and, by virtue of its influence, belongs to an order of phenomena on a cosmic-planetary scale.

If one adds the surface area of the vegetation on land to that of phytoplankton in oceanic water column, the resulting sum represents a surface area vaster than the ocean itself. In fact, the photosynthesizers of Earth can be shown to have approximately the same surface area as Jupiter — 6.3×10 square kilometers.[*]

It is no coincidence that the surface area of the biosphere, an entity of cosmic scale, rivals that of the other major objects in the solar system.

The Earth's surface area is a little less than 0.01 percent (0.0086%) of that of the sun, whereas the photosynthesizing surface area of the biosphere is of an altogether different order: 0.86 to 4.2 percent of the sun's surface.[133]

57 These figures, obviously, correspond approximately to the fraction of solar energy collected by living green matter in the biosphere. This coincidence might serve as a departure point from which we can begin to explain the verdure of the Earth.

The solar energy absorbed by organisms is only a small part of what falls on the Earth's surface, and the latter is an insignificant fraction of the sun's total radiation. According to S. Arrhenius,[134] the Earth receives from the sun 1.66×10^{21} kilocalories per year, while the sun emits 4×10^{30}.

This is the only cosmic energy we can consider in our present state of knowledge. The total radiation that reaches the Earth from all stars is probably less than 3.1×10^{-5} percent of that from the sun, as I. Newton demonstrated.[135] The energy from the planets and the moon, mostly reflected solar radiation, is less than one ten-thousandth of the total from the sun.

[*] This assumes that 5 percent of Earth's surface is devoid of green vegetation, and that the green, absorbing surface is increased by a factor of 100 to 500 by multiplication. The maximum green area then corresponds to 5.1×10^{10} to 2.55×10^{11} square kilometers.

133 Vernadsky's figure here is unrealistically high; the surface area of Earth's vegetation is approximately equal to the planetary surface area (Schlesinger, 1991).

134 See Arrhenius, 1896.

135 See Newton, 1989 .

A considerable part of all the incoming energy is absorbed by the atmosphere and only 40 percent (6.7×10^{20} calories per year) actually reaches the surface. This is available for green vegetation, but most of it goes into thermal processes in the crust, the ocean, and the atmosphere. Living matter also absorbs a considerable amount in the form of heat which, while playing an immense role in the sustenance of life, does not directly participate in the creation of the new chemical compounds that represent the chemical work of life.

For chemical work, i.e., the creation of organic compounds that are unstable in the thermodynamic field of the biosphere (§89), green vegetation uses primarily wavelengths between 670 and 735 nanometers (Dangeard and Desroche, 1910-1911[136]). Other portions of the visible part of the electromagnetic spectrum, at wavelengths between 300 and 770 nanometers, are also utilized by green plants to power photosynthesis, but are not used as intensely as those within the 670-735 range. The fact that green plants make use of only a small part of the solar radiation that falls on them is related to the requirements of the chemical work required, rather than to imperfections in the transforming apparatus.

According to J. Boussingault,[137] one percent of the solar energy received by a cultivated green field may be used for conversion of energy into organic, combustible[138] matter. S. Arrhenius[139] calculates that this figure may reach two percent in areas of intense cultivation. H. T. Brown and F. Escombe[140] found by direct observation that it reaches 0.72 percent for green leaves. Forest-covered surfaces make use of barely 0.33 percent, according to calculations based upon woody tissue.

58 These are undoubtedly minimum figures. In Boussingault's calculation,[141] which included Arrhenius' correction,[142] only vegetation on land was considered. It should be assumed, moreover, that the fertility of the soil is increased by cultivation, and that the favorable conditions we create apply not only for valuable cultivated plants, but also for weeds. These calculations do not account for the lives of the weeds and the microscopic photosynthesizers that benefit from the favorable conditions provided by cultivation and manure. The Earth also has rich concentrations of life other than fields, such as marshes, humid forests, and prairies, where the quantity of life is higher than in human plantings. (§150 et seq.)

The principal mass of green vegetation is in the oceans,[143] where the animal world assimilates vegetable matter as fast as

136 See Dangeard, 1910a, b, 1911a-d; and Desroche, 1911a-e.

137 See Boussingault, 1860-84.

138 It is likely that Vernadsky used the terms "organic" and "combustible" as synonyms.

139 See Arrhenius, 1896.

140 See Brown and Escombe, 1898, 1900.

141 See Boussingault, 1860-84.

142 See Arrhenius, 1896.

143 As noted earlier, this statement is incorrect.

the latter is produced. The rate of production depends upon the quantity of green life per unit area, and appears to be about the same as on land. The collections of heterotrophic animal life, which are thus formed in the plankton and benthos of the ocean, occur on a scale that can rarely, if ever, be seen on land.

We have mentioned that the minimum figure of Arrhenius[144] must be increased, and it should be noted that a correction of the order indicated by this author is already apparent.

Green matter absorbs and utilizes, it would appear, up to 2 percent or more of radiant solar energy. This figure falls within the limits 0.8 to 4.2 percent, which we calculated as the fraction of the solar surface which would have an area equal to the green transforming surface of the biosphere (§56). Since green plants have at their disposal only 40 percent of the total solar energy reaching the planet, the 2 percent that they use corresponds to 0.8 percent of the total solar energy.

59 This coincidence can be explained only by admitting the existence of an apparatus, in the mechanism of the biosphere, that completely utilizes a definite part of the solar spectrum. The terrestrial transforming surface created by the energy of radiation will correspond to the fraction of total solar energy that lies in the spectral regions capable of producing chemical work on Earth.

We can represent the radiating surface of the rotating sun that lights our planet by a luminous, flat surface of length AB (Fig. 1). Luminous vibrations are constantly directed to the Earth from each point of this surface. Only a few hundredths of m percent of these waves, having proper wavelengths, can be converted by green living matter into active chemical energy in the biosphere.

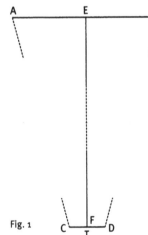

Fig. 1

The rotating surface of the Earth can also be represented, by a plane surface illuminated by solar rays. Considering the enormous size of the solar diameter, and the distance from the Earth to the sun, this surface can be expressed in the figure by the point T, which may be considered as the focus of the solar rays leaving the luminous surface AB.

The green transmuting apparatus in the biosphere is composed of a very fine layer of organized particles, the chloroplasts. Their action is proportional to their surface, because the chlorophyll itself is quite opaque to the chemically-active frequencies of light it transforms. The maximum transformation of solar energy by green plants will occur when there is a receiver on the Earth having a plane surface at least equal to m percent of the luminous (plane) surface of the sun. In this case, all the rays necessary for the Earth will be absorbed by the chlorophyll-bearing apparatus.

In the illustration, CD represents the diameter of a circle with surface equal to 2 percent of the solar surface;* AB represents the diameter of a circle with surface equal to the whole radiating area of the sun; CD similarly represents the area receptive to radiations falling on the Earth; T corresponds to the surface of the Earth. Unknown relationships probably exist connecting the solar radiation, its character (the fraction m of chemically-active radiations in the biosphere), and the plane surface of green vegetation and of azoic areas. It follows that the cosmic character of the biosphere should have a profound influence on the biota thus formed.

60 Living matter always retains some of the radiant energy it receives, in an amount corresponding to the quantity of organisms. All empirical facts indicate that the quantity of life on the Earth's surface remains unchanged not only during short periods, but that it has undergone practically no modification[†] throughout geological periods from the Archean to our own times.

The fact that living matter is formed by radiant energy lends great importance to the empirical generalization regarding the constancy of the mass of living matter in the biosphere, since it forms a connection with an astronomical phenomenon; namely, the intensity of solar radiation. No deviations of any importance in this intensity throughout geological time can be verified. When one considers the connection between green living matter — the principal element of life — and solar radiations of certain wavelengths, as well as our perception that the mechanism of the biosphere is adapted to complete utilization of such rays by green vegetation, we find a fresh and independent indication of the constancy of the quantity of living matter in the biosphere.

61 The quantity of energy captured every moment in the form of living matter can be calculated. According to S. Arrhenius,[145]

* In the illustration, surfaces are reduced to areas, taking the radius of a circle having an area equal to that of the sun as unity.
The radius of the circle having the same area as the sun:
$r = 4.3952 \times 10^6$ kilometers (taken as 1)

The radius of the circle having the same area as the Earth:
$r_1 = 1.2741 \times 10^4$ kilometers (taken as 0.00918)

The radius of the circle having the same area as $0.02 \times$ area of the sun:
$r_2 = 1.9650 \times 10^5$ kilometers (taken as 0.14148)

The radius of the circle having the same area as $0.008 \times$ area of the sun:
$r_3 = 1.2425 \times 10^5$ kilometers (taken as 0.08947)

The mean distance from Earth to sun is 1.4950×10^8 km (taken as 215, relative to the radius of a circle having area equal to that of the sun).

† That is, it oscillates about the stable static state, as in the case of all equilibria.

145 See Arrhenius, 1896.

the combustible, organic compounds produced by green vegetation contain 0.024 percent of the total solar energy reaching the biosphere — 1.6×10^{17} kilocalories, in a one-year period.

Even on a planetary scale, this is a high figure, but it seems to me that it should be even larger than stated. We have tried to show elsewhere[*] that the mass of organic matter calculated by S. Arrhenius,[146] based upon the annual work of the sun, should be increased ten times, and perhaps more. It is probable that more than 0.25 percent of the solar energy collected annually by the biosphere is constantly stored in living matter, in compounds that exist in a special thermodynamic field,[147] different from the field of inert matter in the biosphere.

The quantity of substances constantly moving through life is huge, as illustrated by the production of free oxygen (approximately 1.5×10^{21} grams/year).[148] Even larger, however, is the effect of the creatures that are constantly dying, and being replaced by multiplication. We have seen (§45) that the mass of elements that migrate in the course of a year exceeds, by many times, the weight of the upper 16 kilometers of the Earth's crust (of the order of 10^{25} grams).

As far as can be judged, the energy input to the biosphere, in the course of a year, by living matter does not much exceed the energy that living matter has retained in its thermodynamic field for hundreds of millions of years. This includes at least 1×10^{18} kilocalories in the form of combustible compounds. At least 2 percent of the energy falling on the surface of the earth and oceans is expended in the work of new creation and reconstruction; i.e., at least 1.5×10^{19} kilocalories. Even if later study should increase this figure, its order of magnitude can hardly be different from 10^{19}.

Regarding the quantity of living matter as constant throughout geological time, the energy contained in its combustible part can be regarded as an inherent and constant part of life. A few times 10^{19} kilocalories will thus be the energy transmitted by life, annually, in the biosphere.

[*] See Vernadsky, 1924, p. 308.

146 See Arrhenius, 1896.

147 Such as the cambial wood of trees.

148 Vernadsky had a math (or proofreading) problem with his oxygen values. He uses this same number earler (1.5×10^{21}) to describe the total resevoir of atmospheric oxygen, whereas here he is using it to describe annual generation. The correct value is closer to 2.7×10^{17} grams of O_2/year.

Some Remarks on Living Matter in the Mechanism of the Biosphere

62 Photosynthetic living matter does not include all the essential manifestations of life in the biosphere, because the chemistry of the biosphere is only partially controlled by the vegetable world. Certain regularities that can be regarded as empirical (if not fully understood) generalizations are frequently encountered in nature, and in spite of their uncertainties must be taken into account. The most essential of them are described below.

The eminent naturalist K. E. Baer long ago noted a peculiarity that governs the whole geochemical history of living matter — *the law of economy of utilization* of simple chemical bodies after they have entered into the biosphere. Baer demonstrated this in connection with carbon, and later with nitrogen; it can be extended to the geochemical history of all chemical elements.[149]

Economy in the utilization of chemical elements by living matter is manifested, in the first instance, within organisms themselves. When an element enters an organism, it passes through a long series of states, forming parts of many compounds, before it becomes lost to the organism. In addition, the organism introduces into its system only the required quantities of these elements, avoiding any excesses. It makes choices, seizing some and leaving others, always in a definite proportion.[150] This aspect of the phenomenon to which Baer gave his attention is evidently connected with the autonomy of the organism, and with the systems of equilibrium[151] which enable it to achieve stability and to minimize its free energy.

In larger masses of living matter, this law of economy is demonstrated with even greater clarity. Once atoms become involved in the vital vortices of living matter, they escape only with difficulty into the inert matter of the biosphere, and some perhaps never escape at all. Countless heterogeneous mechanisms absorb atoms into their moving medium, and preserve them by carrying them from one vital vortex to another. These include parasites, organisms which assimilate other organisms, new generations produced by multiplication, symbioses, and saprophytes.[152] The latter make use of the remains of life, much of which are still living because they are impregnated with microscopic forms, transforming them rapidly into a form of living matter.

So it has been, throughout the whole vital cycle, for hundreds

149 See Baer, 1828, 1876.

150 And with a selectivity that can distiguish heavier from lighter isotopes of such elements as oxygen and carbon.

151 American biologists today call this equilibrium *homeostasis*, but there was no such term in Vernadsky's time.

152 One could perhaps carry this thought further and argue that, over geological time, the ratio of "living matter" to "inert matter" in the biosphere has increased, with the forces of life tending to increase the amount of matter in circulation as part of something that is alive. Parasites and hyperparasites have colonized the land-based living environment of their hosts' tissues, environments that (partly as a result of these symbioses) have spread over the surface of Earth.

153 Presumably Vernadsky is referring here back to what he had earlier called "compounds that exist in a special thermodynamic field," such as wood and certain types of animal tissues.

154 Indeed, isotopic fractionation, now a well-known characteristic of life, was first hypothesized in Russia by Vernadksy (1939b). Vernadsky carried his view of life as a geological force to the point of proposing that biogeochemistry of elements in organisms be used as a form of taxonomy of organisms! In the paper cited above (p. 7-8), Vernadsky seems to express some disappointment that the technique is not working out as cleanly and simply as he had hoped:
"Proceeding from this general statement, it has been possible to show by the work of our Laboratory that [emphasis his] the *atomic composition of organisms, plants, and animals is as characteristic a feature as their morphological form or physiological structure or as their appearance and internal structure*. It should be noted that the elementary chemical composition of living organism[s] of the same species taken at different times, in different years, at different places, for instance in Kiev or in Leningrad, varies less than a natural isomorphous mixture of minerals,

of millions of years. A portion of the atoms of the unchangeable covering layer, which possesses a nearly uniform level of energy of about 10^{19} kilocalories, never leaves this vital cycle. As visualized by Baer, life is parsimonious in its expenditure of absorbed matter, and parts from it only with difficulty. Life does not easily relinquish the matter of life, and these atoms remain associated with life for long stretches of time.

63 Because of the law of economy, there can be atoms that have lived in the framework of living matter throughout whole geological periods, moving and migrating, but never returning to the source of inert matter.[153]

The unexpected picture outlined by this empirical generalization forces us to examine its consequences and seek an explanation. We can proceed only hypothetically. To begin with, this generalization raises a question that science has not yet considered, although it has been discussed in philosophical and theological circles: are the atoms which have been absorbed in this way, by living matter, the same as those of inert matter? Or do special isotopic mixtures exist among them?[154] Only experiment can give an answer to this problem, which is of great interest for contemporary science.

64 The exchange of gases between organisms and their surrounding medium is a life process of immense importance in the biosphere (§42). One part of this exchange has been explained by L. Lavoisier[155] as combustion, by means of which atoms of carbon, hydrogen, and oxygen perpetually go and come, inside and outside living vortices.

Combustion probably does not reach the essential substratum of life, the protoplasm. It is possible that the atoms of carbon set free as carbon dioxide by the living organism are derived from matter foreign to the organism, such as food, and not from elements that are part of its framework. If this is so, then the atoms that are absorbed and retained by living matter will collect together only in protoplasm and its structures.[156]

The theory of the atomic stability of protoplasm originated with C. Bernard.[157] Although not accepted by orthodox biologists, it resurfaces from time to time and awakens the interest of scholars. Perhaps a connection exists between Bernard's ideas, Baer's generalization on vital economy,[158] and the empirical fact of the constancy of the quantity of life in the biosphere. All these ideas may be connected with *the invariability of the quantity of*

easily expressed by stoichiometric formulas. The composition of different species of duckweed or insects is more constant than the composition of orthoclases [feldspars] or epidotes [greenish calcsilicate minerals] from different localities. For organisms there is a narrow range within which the composition varies, but there are no stoichiometric[ally] simple ratios for them . . . It may be assumed that in all the cases so far investigated we find a confirmation of the fundamental principle of biogeochemistry, namely, that *numerical biogeochemical features are specific, racial and generic characteristics of the living organisms.* As yet it has been possible to establish it precisely for many species of plants and insects. But it is already clear that this is a general phenomenon. The relations are not so simple as one could have presumed. Many questions evidently arise that require biological criticism."

It is clear from this passage that Vernadsky not only wishes to view life as a geological force, but also individual life forms as minerals. This view continues to influence Russian work on the interpretation of metal contents of various organisms (e. g., Tropin and Zolotukhina, 1994; and Timonin, 1993) and on the ability of microorganisms to mobilize metals and influence the history of mineralogy (Karavaiko, Kuznetsov and Golomzik, 1972; and Kuznetsov, Ivanov and Lyalikova, 1962). Vernadsky's research on the history of minerals of Earth's crust (1959) has generated a unique development of Russian thought on this issue (e. g., A. S. Povarennykh, 1970). Vernadsky's imprint is also apparent in the development of Russian thought on the relationship between the biosphere, granites and ore deposits (see Tauson, 1977).

At least one western scientist has focused on elemental distinctions between taxa (Morowitz, 1968). Morowitz's Table 3-2 recalls Vernadsky's geochemical taxonomy of organisms, and compares the C, H, N, O, P, S, Ca, Na, K, Mg, Cl, Fe, Si, Zn, Rb, Cu, Br, Sn, Mn, I, Al, and Pb contents of man, alfalfa, copepod, and bacteria. Data for the copepod (*Calanus finmarchicus*) were taken from Vernadsky paper, 1933a, p. 91.

155 See Lavoisier, 1892.

protoplasmic formations in the biosphere throughout geological time.

65 The study of life-phenomena on the scale of the biosphere shows that the functions fulfilled by living matter, in its ordered and complex mechanism, are profoundly reflected in the properties and structures of living things.

In this connection, the exchange of gases must be placed in the first rank. There is a close link between breathing and the gaseous exchange of the planet.

J. B. Dumas and J. Boussingault showed,[159] at a remarkable conference in Paris in 1844, that living matter can be taken as an *appendage of the atmosphere*.[160] Living matter builds bodies of organisms out of atmospheric gases such as oxygen, carbon dioxide, and water, together with compounds of nitrogen and sulfur, converting these gases into liquid and solid combustibles that collect the cosmic energy of the sun. After death, it restores these same gaseous elements to the atmosphere by means of life's processes.

This idea accords well with the facts. The firm, generative connection between life and the gases of the biosphere is more profound than it seems at first sight. The gases of the biosphere are generatively linked with living matter which, in turn, determines the essential chemical composition of the atmosphere. We dealt earlier with this phenomenon, in speaking of gaseous exchange in relation to the creation and control of multiplication and the geochemical energy of organisms. (§42)

The gases of the entire atmosphere are in an equilibrium state of dynamic and perpetual exchange with living matter. Gases freed by living matter promptly return to it. They enter into and depart from organisms almost instantaneously. The gaseous current of the biosphere is thus closely connected with photosynthesis, the cosmic energy factor.

66 After destruction of an organism, most of its atoms return immediately to living matter, but a small amount leave the vital process for a long time. This is not accidental. The small percentage is probably constant and unchangeable for each element, and returns to living matter by another path, thousands or millions of years afterwards. During this interval, the compounds set free by living matter play an important role in the history of the biosphere, and even of the entire crust, because a significant fraction of their atoms *leave the biosphere* for extended periods.

156 It is now known, of course, that organisms are able to catabolize both food taken in and biomolecules forming part of the body structure. Vernadsky appears to be arguing here for a permanent sequestering of some atoms in living structure, but this is not generally the case.

157 See Bernard, 1866, 1878.

158 See von Baer, 1876.
159 See Boussingault and Dumas, 1844a abd 1844b. But also see Boussingault and Dumas, 1841, which may be the report to which Vernadsky is referring; in that case Vernadsky is incorrect about the date of the conference.

160 Dumas and Boussingault thus anticipated the Lovelockian view of the intimate relationship between life and atmosphere, matching Lamarck's anticipation of Vernadsky's articulation of full concept of the biosphere.

We now have a new process to consider: *the slow penetration into the Earth of radiant energy from the sun.*[161] By this process, living matter transforms the biosphere and the crust. It constantly secretes part of the elements that pass through it, creating an enormous mass of minerals unique to life; it also penetrates inert matter of the biosphere with the fine powder of its own debris.[162] Living matter uses its cosmic energy to produce modifications in abiogenic compounds (§140 et seq.). Radiant energy, penetrating ever-more-deeply due to the action of living matter on the interior of the planet, has altered the Earth's crust throughout the whole depth accessible to observation. Biogenic minerals converted into phreatic[163] molecular systems have been the instruments of this penetration.

The inert matter of the biosphere is largely the creation of life.[164]

We return, in a new venue, to the ideas of natural philosophers of the early 19th century: L. Ocken,[165] H. Steffens,[166] and J. Lamarck.[167] Obsessed with the primordial importance of life in geological phenomena, these thinkers grasped the history of the Earth's crust more profoundly, and in better accordance with empirical facts, than generations of the strictly observation driven geologists who followed.[168]

It is curious that these effects of life on the matter of the biosphere, particularly on the creation of agglomerations of vadose minerals, are chiefly connected with the activity of aqueous organisms. The constant displacement of aqueous basins, in geological times, spread chemical free energy of cosmic origin throughout the planet. These phenomena appear as a stable dynamic equilibrium, and the masses of matter that play a part in them are as unchanging as the controlling energy of the sun.

67 In short, a considerable amount of matter in the biosphere has been accumulated and united by living organisms, and transformed by the energy of the sun. The weight of the biosphere should amount to some 10^{24} grams.[169] Of this, activated living matter that absorbs cosmic energy accounts for, at most, one percent, and probably only a fraction of one percent. In some places, however, this activated living matter predominates, constituting 25 percent of thin beds such as soil.

The appearance and formation of living matter on our planet is clearly a phenomenon of cosmic character. It is also very clear that living matter becomes manifest without abiogenesis. In other words, living organisms have always sprung from living

161 Preston Cloud (Cloud, 1983, p. 138) defined the biosphere as "a huge metabolic device for the capture, storage and transfer of energy."

162 Vernadsky emphasizes here how the energy from sunlight can penetrate downward into the crude matter of the biosphere as plant roots, deep soil microbes, the hot deep biosphere, etc.

163 Of, or pertaining to, ground water or the zone of water saturation in soil.

164 In his later work Vernadsky considered inert and biogenic matter separately. He defined the inert matter as "the matter created by processes in which living matter does not participate," and the biogenic matter as "matter which has been created and processed by life" (Vernadsky, 1965, pp. 58-60). Vernadsky had not yet made this distinction at the time of the writing of *The Biosphere*. For more details, see Lapo, 1987.

165 See Ocken, 1843.

166 See Steffens, 1801.

167 See Lamarck, 1964.

168 This paragraph relates directly to Vernadsky's main purpose in writing this book. He wanted to demonstrate the primacy of life as a geological force, and to show that life makes geology. Living processes have a fairly direct influence over, even control of, all crustal geological processes. Vernadsky was absolutely correct to emphasize this point, and this is exactly what makes *The Biosphere* such an important book.

169 Grace Osmer (McMenamin and McMenamin, 1994, p. 259) calculates the quantity of water in Hypersea to be 19 cubic kilometers, having a mass of 19 km³ × (10^9 m³/1 km³) × (10^6 cm³/1 m³) = 1.9 × 10^{16} cm³ or grams of water. This value represents a significant fraction of Earth's biomass. Vernadsky's mass of the biosphere (10^{24} grams) is hugely greater (by eight orders of magnitude) because it includes the bio-inert parts of the lithosphere as well as activated living matter. The actual amount of living matter is no

organisms during the whole of geological history;[170] they are all genetically connected; and nowhere can solar radiation be converted into chemical energy independent of a prior, living organism.

We do not know how the extraordinary mechanism of the Earth's crust could have been formed. This mechanism is, and always has been, saturated with life. Although we do not understand the origin of the matter of the biosphere, it is clear that it has been functioning in the same way for billions of years.[171] It is a mystery, just as life itself is a mystery, and constitutes a gap in the framework of our knowledge.

doubt much less a fraction of one percent or two to three orders of magnitude difference than the estimate Vernadsky gives. This only serves to underscore Vernadsky's point about the geochemical, catalytic nature of life in the biosphere.

170 Vernadsky thus implies that any "azoic" period would have been pregeological. Oparin (1957, p. 57-59) had much to say about this, and in the end exonerated Vernadsky for finally agreeing that life could have an origin:

"As a result of prolonged and varied studies of the question, we see that Vernadsky abandoned the untenable position of materialistic dualism [life distinct from other matter] which he previously held. In 1944 he wrote 'in our time the problem cannot be treated as simply as it could during the last century when, it seemed, the problem of spontaneous generation had finally been solved in a negative sense by Louis Pasteur's research.'"

Indeed the picture became more complex with Bernal's suggestion that absorption of organic molecules onto clays, assymetric quartz crystals or other minerals would provide for the concentration of molecules required for life's origin, and would prevent reverse reactions (Young, 1971, p. 371). In 1908 Vernadsky championed directed panspermia as support for the eternity of life:

"By the way, it turns out that the quantity of living matter in the earth's crust is a constant. Then *life* is the same kind of part of the cosmos as energy and matter. In essence, don't all the speculations about the arrival of 'germs' [of life] from other heavenly bodies have basically the same assumptions as [the idea of] the eternity of life." (see Bailes, 1990, p. 123).

Bailes (1990, p. 123) criticizes this passage as evidence of a "mystical strain" in Vernadsky's thought. Such criticisms were also levied by Oparin, who complained that Vernadsky's "theories of perpetual life" (p. 59) were not in accord with the "objective data of modern science." Although Bailes seems confused by Vernadsky's letter to Samoilov, the passage helps us now to clarify Vernadsky's thought in these matters. Vernadsky is in effect characterizing life as not merely a geological force but as a cosmic force on a par with energy and matter. Life for Vernadsky is not an epiphenomenon of matter in an energy stream but a comparably powerful entity in and of itself. Comparatively fragile in any given place and at any given moment, the true force of life is manifest over geologic time.

171 This remarkable passage captures the spirit of Part One of *The Biosphere*. In the same way that a geological system is a time-rock unit (all the rocks deposited during a certain interval of geological time), Earth's crust for Vernadsky becomes a life-rock unit, "saturated with life," and ultimately characterized by this life. Life for Vernadsky is the *sine qua non* of Earth's crust as we know it.

What came before on this planet is, just as it was in Vernadsky's time, still quite unknown. The Russian version of substantive uniformitarianism expressed here by Vernadsky has its own scientific validity. As we will see in Part Two, not only does Vernadsky's view permit Earth a history, it *requires* that the living matter of the biosphere develop (*razvitie*) to more complex states.

PART TWO

The Domain of Life

A greening land,
A shining ether....

F. Tyutchev, 1865

The Biosphere: An Envelope of the Earth

68 In 1875, one of the most eminent geologists of the past century, Prof. E. Suess of Vienna University, introduced the idea of the *biosphere* as a specific, life-saturated envelope of the Earth's crust.[172] Despite the importance of life in the structure of the Earth's crust, this idea has only slowly penetrated scientific thinking, and even today is not much appreciated. This idea, focusing on the ubiquity of life and the continuity of its manifestations, represented a new *empirical generalization* of which Suess could not have seen the full implications. Only as the result of recent scientific discoveries is this beginning clear.

69 The Earth can be divided into two classes of structures: first, great concentric regions that can be called *concenters*; and second, subdivisions of these regions called *envelopes* or *geospheres*.* The chemical and physical properties of these structures vary in a consistent manner, dependent upon distance from the center of the Earth. The biosphere forms the envelope or upper *geosphere* of one of these great concentric regions — the crust.

There are at least three great *concenters: the core of the planet, the sima region, and the crust*. It appears that matter in each of these regions is unable to circulate from one concenter to another, except very slowly, or at certain fixed epochs. Since such migration is not a fact of contemporary geologic history, each region would seem to constitute an isolated, and independent, mechanical system.[173]

The Earth remains in the same thermodynamic conditions for millions of years. Where no influx of active energy from outside has occurred, the mechanical systems of the Earth have surely reached stable, dynamic equilibria. There may well be an inverse relation between the stability of the equilibria, and the degree of influx of outside energy.[174]

70 The chemical composition of the *Earth's core* is clearly different from that of its crust. It is possible that the matter composing the core exists in a particular gaseous state, above the critical point. Unfortunately, in the present stage of science, ideas about the physical state of the planet at great depth are entirely conjectural. It is held that these regions are subject to tens, hundreds, or even thousands of atmospheres of pressure. Similarly, for these regions one typically assumes a prevalence of comparatively heavy free elements or their simple compounds.

172 The concept of the biosphere is present in Jean-Baptiste Lamarck's 1802 book *Hydrogéologie* (1964), but Lamarck was referring only to living matter, not inert matter. E. Suess, in his book *The Origin of the Alps* (1875), coined the term as follows: "One thing seems to be foreign on this large celestial body consisting of spheres, namely, organic life . . . . On the surface of continents it is possible to single out a self-contained biosphere" (Suess, 1875, p. 159). Suess used this term only once and left it undefined.

Vernadsky combines the two usages into his definition of the biosphere, which includes living beings (living matter) and the sediments deposited under their influence (crude matter of the biosphere).

* The word "geosphere" is used by several geologists and geographers in the meaning indicated. J. Murray (1913) and D. N. Soboleff (1926) are examples, all based on the ideas of E. Suess.

173 The fact that Vernadsky was willing to entertain the possibility of matter transfer between the crust, the mantle (Suess's *sima*) and the core suggests a mobilistic view of solid Earth that perhaps would have been sympathetic to the notion of continental drift. We now of course know that Earth's crust has gone through several supercontinental cycles (supercontinents Rodinia, Gondwana, and Pangea; Li et al. 1996).

174 This recognition of thermodynamic stability and dynamic equilibrium of Earth's surface was rediscovered by James Lovelock (1983). Vernadsky's suggestion of an inverse relationship between the stability of the equilibria and the degree of influx of outside energy is a fascinating one that to my knowledge has received no adequate test. Vernadsky implies that increased solar influx would destabilize Earth's climate.

But the physical properties of the Earth's core can be outlined in other ways. For example, it can be imagined that the core is solid, gaseous or viscous; that very high or very low temperatures prevail at the core. Certainly, the core has a distinctly different chemical composition than that of the surface. Since the mean specific gravity of the planet is 5.7, and that of the crust is 2.7, the specific gravity of the core can hardly be less than 8, possibly even 10. It is not unlikely that, as hypothesized, it consists of free iron or iron-nickel alloys.[175]

Seismometric studies definitely indicate that, at a depth of some 2900 kilometers below the surface of the ocean, there is a sudden change in the physical properties of the Earth. It is hypothesized that at this depth, seismic waves meet a different concenter, the metallic core. It is possible, however, to put this boundary at the lesser depth of 1200-1600 kilometers, on the basis of other seismic wave discontinuities.

71 Although the past few years have seen great changes in scientific opinions about this region of the Earth, our present knowledge does not permit precise conclusions. The near future will, no doubt, bring great progress in this regard.

Petrogenetic research and seismic observations suggest that silicate and aluminosilicate rocks occupy a much greater place in the structure of our planet that was formerly imagined. The work[176] of the remarkable Croatian father and son, the Mohorovičić, has called attention to this fact.

72 The second *concenter*, which Suess called the *sima*, seemed to him to be characterized chemically by a preponderance of silicon, magnesium, and oxygen atoms. This region is at least several hundreds, and possibly thousands, of kilometers thick. Five principal atoms — silicon, magnesium, oxygen, iron, and aluminum — appear to be its main constituents, with the heavy iron atoms increasing in frequency with depth.

Perhaps rocks analogous to the basic rocks of the crust, the third concenter, also play a great part in the constitution of the sima region. Several geologists and geophysicists have pointed out that the mechanical properties of these rocks are reminiscent of eclogites.

73 At the upper boundary of the sima is the crust, having a mean thickness of nearly 60 kilometers. This figure has been fairly well established by independent data derived from seis-

175 It is presently believed that the pressure at the center of the core reaches 3.5×10^6 atmospheres, with temperatures of 3000° minimum and 5000 to 8000° maximum. Under such conditions, all elements have radically changed properties. It may be that the electron shells are deformed or destroyed. In any case, the density of matter, whether gas or metal vapor, rises. It is known that even at much lower pressures, elements such as silicon acquire metallic properties. Some investigators therefore conclude that silicon may predominate in the core, though not exclusively, and that the percentage of heavier elements, notably iron, should increase. [This was written in 1967. Our knowledge of conditions at Earth's core have not advanced much since, due to the inacessability of this region. However, the lower mantle (which constitutes more than half of Earth's interior by volume) is known to have a higher iron content than previously thought. Recent experiments show that the lower mantle silicate mineral perovskite can take considerable amounts of iron into its crystal structure (McCammon, 1997; Poirier, 1997) —Mark McMenamin].

176 See Mohorovičić, 1910; Mohorovičić, 1915; and Skoko and Mokrovic, 1982.

mology and studies of specific gravity.

The remarkable isostatic surface of the sima clearly differentiates it from the crust. Magma is homogeneous in all the concentric layers of the sima, changing only as a function of distance from the center of the planet. By contrast, the matter of the Earth's crust is definitely heterogeneous. Therefore, no significant exchange between the sima and the crust can take place.[177]

74 It follows that there are no sources of free energy in the sima capable of reacting with the phenomena of the crust. From the standpoint of these phenomena, the energy of the sima is foreign potential energy, which has never manifested itself on the planet's surface in geologic times. Since we can find no trace of its action, this statement can be taken as a well-established empirical generalization. In other words, we have no facts to show that the sima has not been in a state of complete, and permanent, chemical stasis and stable equilibrium throughout the whole of geologic time.

Two facts confirm this view of the sima: 1. there is no scientifically-established case of matter having been brought up from the deeper regions of the Earth;[178] and 2. there is no evidence of free energy (such as a rise in temperature) inherent in the sima. The free energy, the heat that is conducted from deep regions to the Earth's surface, is not connected with the sima, but rather with the energy of radioactive chemical elements concentrated in the crust.

75 Except for earthquakes, the study of gravity anomalies provides a greater insight into the Earth's interior than that offered by any other measurable surface phenomena. These variations are connected with a particular aspect of the structure of the planet's upper region. Concentrated in this region, and distributed vertically so that the lighter parts compensate for the heavier, are great portions of the crust of different density (1.0 for water to 3.3 for basic rocks). At the theoretical depth of isostatic compensation a surface exists where complete equilibrium of matter and energy prevail; mechanical irregularities and chemical differences should be nil throughout layers at the same depth below this isostatic surface.[179] The surface separates the crust from the sima, and is the boundary between the region of change and the region of stasis.

We noted earlier that the biosphere, the upper envelope of the region of change, produces the changes by drawing on the ener-

177 This is now known not to be the case. Subducting slabs of oceanic crust are carried deep into the Sima; see Nafi Toksöz, Minear and Julian, 1971; and Nafi Toksöz, 1975.

178 Kimberlites are an exception to this statement; see Mitchell, 1986; Nixon, 1973; and Cox, 1978.

179 In a geological situation of isostatic equilibrium, various bodies of rock are arranged higher or lower, based on their respective densities, with respect to their distance from Earth's core. In a sense, crustal blocks float on Earth's mantle, and ride higher or lower depending on whether they are less or more dense, respectively.

gy of the sun. This solar energy is transferred to the depths of the crust, and soon we shall consider how this is accomplished.

In addition to the sun, there is another source of free energy — radioactive matter, which causes slow but powerful disturbances in the equilibria of the Earth's crust. Are there any radioactive atoms in the sima? We do not know; but the thermal properties of the planet would be very different if the quantity of such matter there were of the same order as is found in the crust. So, radioactive matter, one of the sources of the Earth's free energy, either does not exist in the sima or diminishes rapidly as one goes downward from the base of the crust.[180]

76 Our ideas about the physical state of matter in the sima are imprecise. Its temperature does not appear to be very high, but its great pressure produces odd effects. Mohorovicic suggested that the mechanical properties of this matter, at least to the depth of 2000 kilometers, are analogous to solids, but that the pressures are so unimaginably high that our experimentally-based notions of the three states of matter — solid, liquid, and gas — are completely inapplicable. The usual parameters which characterize the differences between these states of matter break down even at the upper boundary of the sima, where the pressures reach 20,000 atmospheres, as Bridgman's 1925 experiments demonstrate.[181]

This matter cannot be crystalline; perhaps it assumes a vitreous or metallic state under the high pressures involved. These are perfectly homogeneous layers in which the pressure increases, and properties change, progressively with depth.

77 The thickness of the isostatic surface is not clearly known. At one time, it was thought to be 110 to 120 kilometers; but more recent and precise estimates are much lower. Its level seems to be variable, depending on location, and its form is slowly being altered by what we call the geologic process — that is to say, the action of free energy in the Earth's crust.

Above the isostatic surface is the great concenter called the crust. This term was originally part of the old hypothesis of a formerly incandescent, and liquid, planet, which supposedly left behind a "crust" of consolidation as it cooled.

Laplace gave this cosmology its fullest expression, and for some time it was quite popular with scholars, who[182] exaggerated its scientific value. It gradually became clear that no trace whatever of such a primary crust is to be found in any accessi-

180 This is a very shrewd inference on Vernadsky's part, for there are indeed lesser amounts of radioactive elements in the mantle and core. During Earth's evolution and the differentiation of the crust, radioisotopes such as those of uranium have tended to become concentrated in the crust.

181 See Bridgman, 1925.

182 In particular Lord Kelvin and his calculation of the age of Earth from a molten beginning (Hallam, 1992, p. 124; Kelvin, 1894).

ble geologic strata, and that this hypothetical incandescent-and-liquid past of the planet played no part in geological phenomena. While the hypothesis has been abandoned, the term "crust" continues to be used, but in a different sense.

78 In the Earth's crust, we can distinguish a series of envelopes, the geospheres.[183] Each is characterized by its own dynamic, physical, and chemical equilibria, which are largely isolated and independent. Although the boundaries between them remain difficult for us to establish, and are not usually spherical, the envelopes are arranged concentrically.

The locations of the geospheric boundaries can be determined more precisely in the upper solid and lower gaseous regions of the planet. Large quantities of chemical compounds have reached, and continue to reach, the Earth's surface from depths of 16 to 20 kilometers below sea level, and from 10 to 20 kilometers above. The geologic structure of the Earth shows that the deepest rock masses do not extend below this lower limit. Moreover, the 16 kilometer thickness approximately corresponds to the region containing all sedimentary and metamorphic rocks. The chemical composition of the upper 16-20 kilometers is probably determined by the same geological processes going on today. The general features of this composition are well known.

In spite of the great progress of experimental science, our knowledge becomes less accurate beyond the limits just indicated. Not only are we unable to unambiguously identify the matter which adjoins the Earth's crust, but further, the very states of matter in these regions of high and low pressure are quite unclear.

The only thing that is certain is that knowledge is growing, slowly but surely. The radical revision of old ideas regarding the crust has just begun.[184]

79 We must now turn our attention to several general phenomena which are important to an understanding of the Earth's crust. First, matter in the upper layers of the atmosphere is in a totally different state than what we are accustomed to. The region of the planet above 80-1000 kilometers is a new concenter, in which immense reserves of free energy in the form of electrons and ions are concentrated. The role of this rarefied material medium in the history of the planet is not yet clear.[185]

[183] Vernadsky develops this concept of envelopes further in V. I. Vernadsky, 1942.

[184] Vernadsky's description of the crust precedes the acceptance of plate tectonic theory, but here he seems to be hinting at an awareness of (at the time) newly emerging ideas of continental drift.

[185] This conjecture of Vernadsky's has been decisively confirmed by the discovery of the Van Allen Radiation Belts (see Manahan, 1994, p. 287, fig. 9.9).

Second, the inner layers of the crust are undoubtedly not in a state of liquid incandescence throughout, as was formerly concluded from observations of the eruption of volcanic rocks. It must be admitted that within these layers there are large or small aggregates of magma — masses of silicates in a fused, viscous state — at temperatures of 600° to 1200° C, and dispersed within a primarily solid or semiviscous matrix. There is no indication that these magma bodies penetrate the whole crust, or that they are not concentrated in the upper zone, or that the temperature of the entire crust is as high as that of these incandescent, gas-containing masses.

80 Although the structure of the deeper portions of the crust still harbors enigmas, considerable progress toward understanding this structure has been made in the last few years.

The crust in its entirety seems to consist of the acid and basic rocks found at its surface. Under the continents, to a depth of about 15 kilometers, acid rocks (granites and granodiorites) exist together, but at greater depths the basic rocks predominate. Under the hydrosphere, these basic rocks are poorer in free energy and radioactive elements, and are nearer the surface.

There thus appear to be at least three envelopes beneath the Earth's surface, the upper one corresponding to the acidic, granitic envelope, and extending to a depth of about 15 kilometers, with considerable quantities of radioactive elements.

About 34 kilometers below the surface there is a sudden change in the properties of matter (H. Jeffreys,[186] S. Mohorovicic[187]), probably marking the lower boundary of crystalline bodies, and the upper boundary of R. Daly's "vitreous envelope."[188] Below this level the basic rocks, and in some places the acid rocks, must be in a plastic, perhaps partially molten state that cannot correspond to any rocks known from the Earth's surface.

At the average depth of 59-60 kilometers, there is another abrupt change in the structure of the crust, marked by seismic evidence for dense rock phases, perhaps eclogites* (density not less than 3.3-3.4).

At this level we reach the sima region, in which the density of rocks becomes greater and greater, reaching 4.3 to 4.4 at its base.[189]

81 It has taken many years of empirical work to establish the existence of terrestrial envelopes, although some, such as the atmosphere, have been familiar for centuries.

[186] See Jeffreys, 1924.

[187] See Mohorovičić, 1910; S. Mohorovičić, 1915; and Skoko and Mokrovic, 1982.

[188] See Daly, 1928; Daly, 1938; and Daly, 1940.

* The eclogites are certainly not those known by petrographers, since their structure does not appear to be crystalline, but they have the same density. The eclogites found in the upper portions of the Earth's crust represent the deepest parts of the crust that can be studied visually.

[189] See Williamson and Adams, 1925.

Around the turn of the century we began to grasp the principles of their origin, without fully recognizing their role in the structure of the crust. Their origin and existence are closely connected with chemical processes and laws of equilibrium.

More recently, it has become possible to perceive the complicated chemical and physical structure of the crust, and form at least a simple model of the phenomena and systems of equilibrium that apply to terrestrial envelopes.

The laws of equilibrium were set out in mathematical form by J. W. Gibbs (1884-1887), who applied them to relationships between independent variables in physical and chemical processes (temperature, pressure, physical state, and chemical composition). All of the empirically-recognized geospheres can be distinguished by the variables of Gibbs' equilibria.[190] We can distinguish *thermodynamic envelopes*, determined by values of temperature and pressure; *envelopes of states of matter*, characterized by material phases (solid, liquid, etc.); and *chemical envelopes*, distinguished by chemical composition.

Only the envelope proposed by Suess, *the biosphere*, is left out of this scheme. Its reactions are subject to the laws of equilibrium, but are distinguished by a new property, an independent variable which Gibbs failed to take into account.[191]

82 The above independent variables are not the only ones theoretically possible in heterogeneous equilibria. Electrodynamic equilibria, for example, were studied by Gibbs. In addition, various superficial or electrostatic forces — forces of contact — have a great importance in natural terrestrial equilibria.

In photosynthesis, the independent variable is energy. In phenomena of crystallization, we encounter vectorial energy and internal energy in the formation of crystal twins. Surface energy plays a role in all crystallization processes.

Living organisms bring solar energy into the physico-chemical processes of the crust, but they are essentially different from all other independent variables of the biosphere, because they are independent of the secondary systems of equilibria within the primary thermodynamic field. The autonomy of living organisms is shown by the fact that the parameters of their own thermodynamic fields are absolutely different from those observed elsewhere in the biosphere. For example, some organisms maintain their own individual temperatures (independent of the temperature of the surrounding medium) and have their own internal pressure. They are isolated in the biosphere.

190 See Gibbs, 1902.

191 At the time Vernadsky was writing *The Biosphere*, many scientists expected that by looking at biology one would discover new laws of physics and chemistry. Great surprise was expressed when, with the elucidation of the structure of DNA, the "trick behind it" turned out to be so simple (Judson, 1979, p. 60).

Although the thermodynamic field of the latter determines the regions in which their autonomous systems can exist, it does not determine their internal field.

Their autonomy is also shown by their chemical compounds, most of which cannot be formed outside themselves in the inert milieu of the biosphere. Unstable in this medium, these compounds decompose, passing into other bodies where they disturb the equilibrium and thereby become a source of free energy in the biosphere.

The conditions under which these chemical compounds are formed, in living beings, are often very different from those in the biosphere. In the biosphere, we never observe the decomposition[192] of carbon dioxide and water, for example, although this is a fundamental biochemical process. In our planet this process only takes place in the deep regions of the magmasphere apart from the biosphere, and can be reproduced in the laboratory only at temperatures much higher than those in the biosphere. It is a fundamental observation that living organisms, carriers of the solar energy that created them, may be empirically described as particular thermodynamic fields foreign to, and isolated within, the biosphere, of which they constitute a comparatively insignificant fraction. The sizes of organisms range from 10^{-12} to 10^8 centimeters. Whatever explanation may be given for their existence and formation, it is a fact that all chemical equilibria in the medium of the biosphere are changed by their presence, while the general laws of equilibria of course remain unchangeable. The activity of the sum total of living creatures, or in other words, living matter, is fully analogous to the activity of other independent variables. Living matter may be regarded as a special kind of independent variable in the energetic budget of the planet.

83 The activity of living creatures is closely connected with their food, breathing, destruction and death; or, in other words, with the vital processes by which chemical elements enter and leave them.

When chemical elements enter living organisms, they encounter a medium unlike any other on our planet. We can regard the penetration of chemical elements into living matter as a new *mode of occurrence*.

Their history in this new mode of occurrence is clearly distinct from that to which they are subjected in other parts of our planet, probably because a profound change of atomic systems

192 Vernadsky means here abiogenic decomposition of water and carbon dioxide, both of which are quite stable at Earth's surface. Both molecules, of course, are split in the most common type of photosynthesis (Photosystem II).

occurs in living matter. It may be that the usual mixtures of isotopes do not exist in living matter — a point for experimentalists to decide.[193]

It was formerly thought (and this opinion has not lost all its adherents) that the special history of elemental constituents of living matter could be explained by the predominance of *colloids* in living organisms. But in numerous examples of non-living colloidal systems, what we know of the biochemical history of elements is not in accord with this idea.

The properties of dispersed systems of matter (colloids) are governed by molecules, and not by atoms. This fact alone should dissuade us from looking to colloidal phenomena for the explanation of modes of occurrence, which are always characterized by the states of atoms.[194]

84 In 1921, we advanced the idea of mode of occurrence of chemical elements as a purely empirical generalization.

The occurrence and history of the elements may be classified in different thermodynamic fields, or in definite parts of them. A larger number of modes of occurrence may exist than are observed in the thermodynamic fields of our planet. The states attributed to atoms in stellar systems, for example, in order to explain their spectra, are impossible on Earth (the ionized atoms of Saha-Meg-Nad[195]). In certain stars, these atoms are endowed with an enormous mass, which can be explained only by assuming densities of thousands and tens of thousands of grams per cubic centimeter (A. Eddington).* The states of these stellar atoms obviously have modes of occurrence that are unknown in the Earth's crust. Other modes of occurrence not found on our planet should be observed in the sun's corona (electron gas), in nebulae, in comets, and in the Earth's core.

85 The existence of chemical elements in living matter should be regarded as their particular mode of occurrence, because living organisms are special thermodynamic fields of the biosphere, which profoundly alter the history of the chemical elements in them. This mode clearly should be classed with the other modes in the crust which are characterized by the particular states of their atoms. It can be expected that future research will clarify the modifications which atomic systems undergo, when they enter living matter.

The different modes of occurrence of atoms in the crust are empirically characterized by: 1. a thermodynamic field, specific

193 Isotopic fractionation has indeed been demonstrated in organisms. Organisms do this (with isotopes of oxygen, carbon, etc.) because it is more energetically efficient, in a biochemical sense, to use the lighter isotopes of a given element. See Vernadsky, 1931.

194 This statement typifies Vernadsky's fixation on the elemental makeup of life, as opposed to its molecular makeup.
This probably represents a desire on Vernadsky's part to have study of the biosphere closely linked to the famous periodic table constructed by Vernadsky's professor D. I. Mendeleev (Yanshin and Yanshina, 1988, p. 283; Fersman, 1946).

195 See Saha-Meg-Nad, 1927.

* The matter of the star Sirius B has a density of 53,000. According to the dynamic ideas of N. Bohr and E. Rutherford (although their models are only approximations of reality), electrons are closer to the nucleus of the atoms in this star than in the case of ordinary atoms (F. Tirring, 1925). The displacement observed in the red spectrum of Sirius B, and in spectral lines of bodies of similar density, confirm such enormous densities, according to relativity theory.

for each mode; 2. a particular atomic configuration; 3. a specific geochemical history of the element's migration; and 4. relationships, often unique to the given mode, between atoms of different chemical elements (paragenesis[196]).

86 We can distinguish four modes of occurrence through which the elements of the crust pass, in the course of their history: 1. *Massive rocks and minerals*, consisting primarily of stable molecules and crystals in immobile combinations of elements; 2. *Magmas*, viscous mixtures of gases and liquids containing neither the molecules nor crystals of our familiar chemistry,* but a mobile mixture of dissociated atomic systems; 3. *Dispersed* elements in a free state, separate from one another and sometimes ionized,† suggesting the radiant matter of M. Faraday and W. Crookes; 4. *Living matter*, in which the atoms are generally believed to occur as molecules, dissociated ionic systems, and dispersed modes, although these seem insufficient to explain the empirical facts. It is very likely that isotopes (§83) and the symmetry of atoms also play roles in the living organism which have not yet been elucidated[197].

87 The modes of occurrence of chemical elements play a role in multivariate equilibria similar to those of independent variables such as temperature, pressure, chemical composition, and states of matter. Such modes of occurrence also characterize geospheres reified as the thermodynamic and other envelopes of the crust, as already described (§81). These geospheres can be called *paragenetic envelopes*, because they determine the main features of paragenesis of elements — the laws of their simultaneous occurrence.

The biosphere is the most accessible and the best known of the paragenetic envelopes.

88 The concept of the Earth's crust as a structure of thermodynamic envelopes, of chemical and paragenetic states of matter, is a typical empirical generalization. It has not yet been linked to any theories of geogenesis or to any conception of the universe, and is therefore still not fully explained.[198]

From all that has been said, it can be seen that such a structure is the result of the mutual action of cosmic forces on the one hand; and of the interplay between the matter and the energy of our planet, on the other. This follows from the fact that the character of matter (the quantitative relationships of the elements,

196 Paragenesis is a geological term for the order of formation of associated minerals in a time succession, as often encountered in metamorphic mineral suites.

* Glasses under high temperature and pressure (§80) can be taken as special magmas; possibly they correspond to a new mode of occurrence of chemical elements.

† The ionized state may, perhaps, be a distinct mode of occurrence.

197 A. I. Perelman notes that there are indications that the spin (right or left) of elementary particles plays a significant role in the existence of asymmetry in organisms (Gardner, 1963). Vernadsky was fascinated with such possibilities, and felt that there must be a basic assymetry or anisotropy to the structure of the universe (Mitropolsky and Kratko 1988). This idea has resurfaced recently with the controversial proposal that the universe is anisotropic and has an axis (Nodland and Ralston, 1997; Glanz, 1997; Cowen, 1997).

198 Many theories and hypotheses have appeared in recent times concerning the origin of layers in the crust. A. P. Vinogradov has developed a model in which the formation of the crust from the mantle is viewed as a result of so-called "zonal melting" –A. I. Perelman.

THE PROPERTIES OF THE EARTH'S ENVELOPES

I THERMODYNAMIC ENVELOPES

1 *Upper Envelope* Region of very low pressure and low temperature: 15-600 km (upper 100 km may be a different planetary region.)

2 *Surface Envelope* Pressure about 1 atmosphere: temperature in range -50° to +50°C.

3 *Upper Metamorphic Envelope* (region of cementation) Temperature still below critical temperature of water; pressure does not radically alter solid state.

4 *Lower Metamorphic Envelope* (region of anamorphism) Temperature above critical temperature of water; pressure makes matter plastic.

5 *Magmasphere* Temperature has not yet reached critical condition for all matter (?); limit of Earth's crust (?).

6 *Barisphere* Temperature has reached critical condition for all matter (?)

II GASEOUS ENVELOPES

1 *Upper Stratosphere* Rarefied gases; ions; electrons above 80-100 km.

2 *Stratosphere* Rarefied gases, merging with troposphere at the bottom: above 10-15 km.

3 *Troposphere* Gas under usual conditions 0 to 10-15 km.

4 *Liquid Hydrosphere* 0-3.8 km.

5 *Solid Lithosphere* Characterized by crystalline state of matter.

6 *Viteous Lithosphere* Solid crystalline state absent because of high temperature and pressure; plastic glass full of gases.

7 *Magmatic* Viscous liquid filled with gas in hot solid surroundings (?).

8 *Gas Under High Pressure (?)* Supercritical gas (?)

III CHEMICAL ENVELOPES

1 *Hydrogen (?)* Perhaps particles of "solid" nitrogen.

2 *Helium (?)* 100-200 km.

3 *Nitrogen (?)* Above 70 km (?).

4 *Nitrogen-oxygen* The atmosphere.

5 *Hydrosphere* 0-3.8 km.

6 *Superficially Altered Crust* Characterized by free oxygen, water, and carbon dioxide.

7 *Sedimentary Envelope* Ancient superficially altered crust; 5 km and below.

8 *Granite Envelope* Paragneiss and orthogneiss.

9 *Basaltic*

10 *Silicon-Iron (?)*

IV PARAGENETIC ENVELOPES

1 *Atomic Envelope* Regions of dispersed elements; free atoms in stable form.

2 *Gaseous Envelope* Formed of molecules and atoms (?).

3 *Biosphere* Region of life and colloids.

4 *Region of Molecules and Crystals*

5 *Magmatic Envelope* Solid chemical compounds absent; filled with gases.

V RADIATION ENVELOPES

1 *Electron Envelope*

2 *Ultraviolet Envelope* Short wave radiation and penetrating cosmic rays; radioactive emanations

3 *Light Ray Envelope* Light and heat rays and radioactive emanations.

4 *Heat Ray and Radioactive Envelope* Various radiations, mostly radioactive.

5 *Heat Ray Envelope* Radioactive processes are absent

for example) is neither accidental nor connected solely with geological causes.

This empirical generalization is set out in tabular form in Table I[199] and will serve as a basis for our further discussion.

Like all empirical generalizations, this table must be regarded as a first approximation to reality, susceptible to modification and subsequent completion. Its significance depends upon the factual, empirical material underlying it, and as a result of gaps in our knowledge is very uneven in terms of its significance.

Our knowledge of the first thermodynamic envelope, and other upper envelopes corresponding to it, is founded on a relatively small number of facts, conjectures, and interpolations, which by their nature are foreign to empirical generalization. The same can be said of the fifth thermodynamic envelope and the regions below it. Our knowledge of these domains is, therefore, unreliable, and will be drastically modified as science progresses. Hopefully, new discoveries will lead to radical changes in current opinion in the near future.

In most cases, it is impossible to indicate the precise line of demarcation between envelopes. The surfaces separating them change with time, sometimes rapidly, according to all indications. Their form is complex and unstable.[*][200] Our lack of knowledge regarding the boundary zones portrayed on the table has little importance for the problems considered here, however, because the entire biosphere lies between these layers, within a region of the table filled with statements based upon an enormous collection of facts and free from hypotheses and interpolations.

89 Temperature and pressure are particularly important factors in chemical equilibria, because they apply to all states, chemical combinations, and modes of occurrence of matter. The thermodynamic envelopes corresponding to them are likewise of special importance. Our model of the cosmos always must have a thermodynamic component.

The origin of elements and their geochemical history must, therefore, be classified according to different thermodynamic envelopes. From now on, the term *vadose* will be given to the phenomena and the bodies in the second thermodynamic envelope; *phreatic*, to those of the third and fourth (metamorphic) envelopes; and *juvenile*, to those of the fifth envelope.

Matter confined to the first and sixth envelopes either does not enter into the biosphere or has not been observed in it.

199 Vernadsky later revised this table (Vernadsky, 1954, pp. 66-68.)

* The basalt envelope is below the oceans, probably at a depth close to 10 km for the Pacific Ocean, and deeper for the Atlantic. It is sometimes thought that the granite envelope under the continents is very thick (more than 50 km under Europe and Asia, according to Gutenberg, 1924; and Gutenberg, 1925).

200 Current values place the thickness of oceanic crust at about 5-10 kilometers and continental crust at about 35 kilometers (see Press and Siever, 1982)

Living Matter of the First and Second Orders in the Biosphere

90 While the boundaries of the biosphere are primarily determined by the *field of vital existence*, there is no doubt that a *field of vital stability* extends beyond its boundaries.[201] We do not know how far beyond the confines of the biosphere it can go because of uncertainties about adaptation, which is obviously a function of time, and manifests itself in the biosphere in strict relation to how many millions of years an organism has existed. Since we do not have such lengths of time at our disposal and are currently unable to compensate for them in our experiments, we cannot accurately assess the adaptive power of organisms.[202]

All experiments on living organisms have been made on bodies which have adapted to surrounding conditions during the course of immeasurable time* and have developed the matter and structure necessary for life. Their matter is modified as it passes through geologic time, but we do not know the extent of the changes and cannot deduce them from their chemical characteristics.†

In spite of the fact that the study of nature shows unambiguous evidence of the adaptation of life and the development of different forms of organisms, modified to ensure their continued existence throughout centuries, it can be deduced from preceding remarks that life in the crust occupies a smaller part of the envelopes than is potentially for it to expand into.

The synthesis of the age old study of nature — the unconscious empirical generalization upon which our knowledge and scientific labor rests — could not be formulated better than by saying that life has encompassed the biosphere by slow and gradual adaptation, and that this process has not yet attained its zenith (§112, 122). The pressure of life is felt as an expansion of the field of vital existence beyond the field of vital stability.

The field of vital stability therefore is the product of adaptation throughout time. It is neither permanent nor unchangeable, and its present limits cannot clearly predict its potential limits.

The study of paleontology and ecology shows that this field has gradually increased during the existence of the planet.

91 The field of existence of living organisms is not determined solely by the properties of their matter, the properties of the environment, or the adaptation of organisms. A characteristic role is also played by the respiration and feeding of organisms,

---

201 Weyl (1966) developed this same concept in terms of what he called the "life boundary."

202 See the discussion of the adaptive power of organisms in Cuénot, 1925. Recent experiments may in fact give us tools with which to accurately assess the adaptive power of organisms (Losos et al. 1997; Case, 1997).

* "Immeasurable time" is an anthropocentric notion. In reality, there are laws which have yet to be established, regarding a definite duration of the evolution [probably measured in billions of years] of living matter in the biosphere.

† The limits of life are often sought in the chemical and physical properties of the chemistry of living organisms, such as in proteins that coagulate at 60°-70° C. But the complex capacities for adaptation by organisms must be taken into account. Certain proteins in a dry state do not change at the temperature 100° C (M. E. Chevreul, 1824).

through which the organisms actively select the materials necessary for life.

We have already see the importance of respiration (the exchange of gases) in the establishment of energy systems of organisms, and also of the gaseous systems of the whole planet, especially its biosphere.

This exchange, as well as the transfer of solid and liquid matter from the surrounding environment into the autonomous field of organisms (§82) — feeding — determines their habitat.

We have already discussed this, in noting the absorption and transformation of solar energy by green organisms (§42). We now return to this discussion in more detail.

Of primary importance is identifying the source from which organisms derive the matter necessary for life. From this point of view, organisms are clearly divided into two distinct groups: *living matter of the first order*[203] — autotrophic organisms, independent of other organisms for their food; and *living matter of the second order* — the *heterotrophs* and *mixotrophs*. This distribution of organisms into three groups according to their food intake was proposed by the German physiologist W. Pfeffer[204] in 1880-1890, and is an empirical generalization with rich implications. It is more important to the study of nature than is generally thought.

Autotrophic organisms build their bodies exclusively from inert, non-living matter. Their essential mass is composed of organic compounds containing nitrogen, oxygen, carbon, and hydrogen — all derived from the mineral world. Autotrophs transform this raw material into the complex organic compounds which are necessary for life. The preliminary labors of autotrophs are ultimately necessary for the existence of heterotrophs, which obtain their carbon and nitrogen largely from living matter.

In mixotrophic organisms, the sources of carbon and nitrogen are mixed — these nutrients are derived partly from living matter, and partly from inert matter.

92 The source from which organisms obtain the matter needed for life is more complicated than appears at first sight; nevertheless, the classification proposed by W. Pfeffer seems to state one of the fundamental principles of living nature.

All organisms are connected to inert matter through respiration and feeding, even if only partially or indirectly so. The distinction between autotrophs and heterotrophs is based on

203 Vernadsky uses "order" here in the sense of "trophic level."

204 See Pfeffer, 1881.

autotrophs' independence from other living matter insofar as *chemical elements* are concerned; they can obtain all such elements from their inanimate surroundings.

But in the biosphere a large number of the molecules necessary for life are themselves the products of life, and would not be found in a lifeless, inert medium.[205] Examples are free oxygen, O_2, and biogenic gases such as CO_2, NH_3, H_2S, etc. The role that life plays in the production of natural aqueous solutions is just as important. Natural water (in contrast to distilled water) is as necessary to life as is the exchange of gases.

The limits to the independence of autotrophic organisms should be emphasized as further evidence of the profound effect of life on the chemical character of the inert water in which it exists. We do not have a right to make the facile conclusion that the autotrophic organisms of today could exist in isolation on our planet.[206] Not only have they been bred from other, similar autotrophs, but they have obtained the elements they need from forms of inert matter produced by other organisms.

93 Thus, free oxygen is necessary for the existence of autotrophic green organisms.[207] They themselves create it from water and carbon dioxide. It is a biochemical product which is foreign to the inert matter of the biosphere.

Furthermore, it cannot be proved that free oxygen is the only material necessary for life that originates from life itself. For example, J. Bottomley has raised the problem of the importance of complex organic compounds for the existence of the green aquatic plants called auximones.[208] The actual existence of auxonomes has not been established, and the hypothesis has been questioned, but Bottomley's research has touched upon a more general question. In the scientific picture of nature, the importance of barely detectable traces of organic compounds present in natural fresh or salt water is becoming more and more apparent. The reservoir of organic matter in the biosphere has a total mass of at least several quadrillion tons. It cannot be said that this mass originates exclusively with autotrophic organisms. On the contrary, we see at every step the immense importance of the nitrogen-rich compounds created by the heterotrophic and mixotrophic organisms. These are especially important in the life of organisms as food, and in the origin of carbon-rich mineral deposits.

Nature continually makes these materials visible to the naked eye without recourse to chemical analysis. They form fresh

205 For Vernadsky, biogenic macromolecules are undoubtedly "products of life." This partly explains his difficulties with concepts of abiogenesis, difficulties which still afflict the study of the origin of life. The fundamental problem is of the chicken and egg variety—how can life be created from macromolecules which themselves are products of life? (See Cairns-Smith, 1991).

206 Vernadsky does not like to take organisms out of their biospheric and historical context. Vernadsky might
have objected to the thought-based experiment, discussed by J. W. Schopf (personal communication), that one cyanobacterium, innoculated into a sterile Earth, could (with unrestricted growth) oxygenate the atmosphere in forty days.

207 Vernadsky seems to be unaware of the acute toxicity of oxygen to unprotected organisms. This may also explain his apparent rejection of Oparin's suggestion of an early reducing atmosphere on Earth. Current research suggests that the early atmosphere was indeed anoxic, but was not as reducing as proposed by Oparin and Haldane. Rather, it was a carbon dioxide rich atmosphere (see Schwartzman and Volk, 1989).

208 See Bottomley, 1917

water and salt water foams, and the iridescent films which completely cover aquatic surfaces for thousands and millions of square kilometers. They color rivers, marshy lakes, tundras, and the black and brown rivers of tropical and subtropical regions. No organism is isolated from these organic compounds, even if they are buried within the earth, because these compounds are continually penetrating it by means of rain, dew, and solutions of the soil.

The quantity of dissolved and colloidal organic matter in natural water varies between 10^{-6} and 10^{-2} percent, and has a gross mass amounting to 10^{18} to 10^{20} tons, mostly in the ocean. This quantity is apparently greater than that of living matter itself.[209] The idea of its importance is slowly entering contemporary scientific thinking. Even with the old naturalists, we occasionally come across an interpretation of this impressive phenomenon, sometimes from an unexpected point of view.

During the years 1870-1880, the gifted naturalist J. Mayer briefly pointed out the important role of such matter in the composition of medicinal waters and in the general economy of nature.[210] This role is even deeper and more striking than he supposed with regard to the origin of vadose and phreatic minerals.

94 The fact that some of the compounds of inert matter required by autotrophic organisms has a biochemical origin does not diminish the distinction between them and heterotrophic or mixotrophic organisms. This can be made clear by giving a somewhat more restricted definition of autotrophy, and adhering to it in our future discussions.

We will call "autotrophs" those organisms of the biosphere which draw the elements necessary for their sustenance from the surrounding inert matter of the biosphere in which they exist, and which have no bodily need for organic compounds prepared by other living organisms.

Such a laconic definition cannot adequately cover this entire phenomenon, since there must necessarily be transitional states and borderline cases, such as the saprophytes which feed on dead and decomposing organisms. The essential food of saprophytes, however, is nearly always composed of microscopic living beings which have entered the remains of dead organisms.

In considering the idea that autotrophic organisms are limited to the current biosphere, we exclude the possibility of drawing conclusions about the Earth's past, and particularly about

209 According to E. A. Romankevich (1984), the average content of dissolved organic carbon in oceanic water is 1.36×10^{-4} %, for a total mass of approximately 1.9×10^{12} tons. The actual value is not well constrained but probably lies between Vernadsky's and Romankevich's values (Sugimura and Suzuki, 1988).

210 Vernadsky's reference could not be located, but see Mayer, 1855.

the possibility that life began on Earth in the form of some kind of autotrophic organisms; for it is certain that the presence of vital products in the biosphere is indispensable for all existing autotrophic organisms.[211] (§92)

95 The distinction between primary producers and consumers is shown by their distribution in the biosphere. The regions accessible to primary producers with autotrophic lifestyles are always more extensive than the habitat of organisms which must consume living matter.

Autotrophs belong to one of two distinct groups: photosynthetic plants and autotrophic bacteria. The latter are characterized by their minute dimensions and their great powers of reproduction.

We have already seen that photosynthesizing organisms are the essential mechanism of the biosphere, and the source of the active chemical energy, both of the biosphere and, to a great extent, the entire crust.

The field of existence of these green autotrophic organisms is primarily determined by the domain of solar radiation. Their mass is very large, compared to the mass of live animals; it is perhaps half[212] of all living matter. Some of these organisms are adapted to the capture and full utilization of feeble luminous radiations.

It has been argued, that at various times in the Earth's past, the extent of photosynthetic life was greater or lesser than it is now, and indeed this may very well have be the case, although it is not yet possible to be certain about it.

The immense quantity of matter contained in green organisms, together with their ubiquity and their presence wherever solar radiation reaches, sometimes gives rise to the idea that these photosynthesizers constitute the essential basis of life.

One sees also that in geologic times they have been transformed by evolution into a multiplicity of organisms which constitute second-order living matter.[213] At the present time, they control the fate of both the entire animal world, and also the immense number of other organisms that lack chlorophyll (such as fungi and bacteria).

Green organisms carry out the most important chemical transformation that takes place on the Earth's crust — the creation of free oxygen, by photosynthetic destruction of oxides as stable and as universal as water and carbon dioxide. Photosynthesizers undoubtedly performed this same work in the past

211 Vernadsky's insight here was rediscovered by J. W. Schopf (1978), who notes that the very earliest organisms must have been heterotrophs. Although it might appear that Vernadsky is ensnared here by his insistence on substantive uniformitarianism, in fact he is presenting a sophisticated argument, the idea that our understanding of biopoesis is flawed because it lacks a satisfactory materialistic explanation.

212 Or more.

213 This idea that green plants give rise to "second-order living matter," that is, primary and secondary (and higher order) consumers, was also expressed in 1953 by A. C. Hardy. Few zoologists would endorse this idea today, although "Garden of Ediacara" theory suggests that many of the earliest animals may have been photosymbiotic (McMenamin, 1986). Achlorophyllous plants such as *Monotropa*, of course, did evolve from green plants.

periods of geologic history. The phenomena of superficial weathering clearly show that free oxygen played the same role in the Archean Era that it plays now in the biosphere. The composition of the products of superficial weathering, and the quantitative relationships that can be established between them, were the same in the Archean Era as they are today. The realm of photosynthesizers in those distant times was the source of free oxygen, the mass of which was of the same order as it is now.[214] The quantity of living green matter, and the energy of solar radiation that gave it birth, could not have been perceptibly different in that strange and distant time from what they are today.[215] (§57)

We do not, however, possess the remains of any photosynthesizers of the Archean Era.[216] These remains only begin to appear continuously from the Paleozoic, but then clearly show intense and uninterrupted evolution of innumerable forms, culminating in some 200,000 species described by biologists. The total number of species which have existed, and which exist today, on our planet is not an accidental one; but it cannot yet be calculated, since the relatively small number of fossil species (several thousands) merely indicates the imperfection of our knowledge. The number of described fossil species continues to grow each year.[217]

96 Autotrophic bacteria represent a smaller quantity of living matter. S. N. Vinogradsky first discovered them at the end of the 19th century,[218] but the concept of organisms lacking chlorophyll and independent of solar radiation has not yet affected scientific thought as it should have. In contrast, the existence of autotrophic green organisms was discovered in the late 18th and early 19th centuries, and their geochemical significance was brought to light by J. Boussingault, J. B. Dumas,[219] and F. V. J. Leibig[220] in the years 1840 to 1850. The role of autotrophic bacteria in the geochemical history of sulfur, iron, nitrogen and carbon is very important; but these bacteria show little variation, in the sense that only about a hundred species are known, and their mass is incomparably less than that of green plants.

These organisms are widely dispersed — in soil, the slime of aqueous basins, and sea water — although nowhere in quantities comparable to autotrophic green plants on land, or oceanic green plankton. The geochemical energy of these bacteria is, however, the highest for any living matter, and higher than that of green plants by a factor of ten or a hundred. The overall kinetic geochemical energy per hectare will be of the same order for both unicellular green algae and bacteria; but while the algae

214 This incorrect line of argument helps explain why Vernadsky missed the connection between oxygenation of the atmosphere and the Precambrian banded iron formations.

215 The composition of Archean erosion products was probably very different from present ones. This was shown, for example, by the Finnish geochemist Rankam, for the products of the lower Bothnian erosion of diorites in Finland. Most investigators now accept that the initial atmosphere of Earth did not contain oxygen. –A. I. Perelman.

216 Now we do: see Awramik, Schopf and Walter, 1983.

217 P. R. Erlich and E. O. Wilson (1991) conjecture (p. 759) that "it is easily possible that the true number [global total] of species is closer to 10^8 than 10^7."

218 S. N. Vinogradsky's discovery of chemoautotrophic microorganisms, a discovery of overwhelming importance for biospheric studies, was made in 1887; see Vinogradsky, 1887. (See also Lapo, 1990, p. 205).

219 See Boussingault and Dumas, 1844.

220 See Leibig, 1847.

can reach the maximum stationary state in about ten days, bacteria in favorable conditions need only a tenth of this time.

221 See J. Reinke, 1901.

222 See Cholodny, 1926.

97 There are only a few recorded observations on the multiplication of autotrophic bacteria. According to J. Reinke,[221] they appear to multiply more slowly than other bacteria; N. G. Cholodny's observations[222] of iron bacteria do not contradict this. These bacteria divide only once or twice in 24 hours ($D = 1$ to 2 per day); ordinary bacteria divide as slowly as this only under unfavorable conditions. For example, *Bacillus ramosus* (which inhabits rivers) yields at least 48 generations in 24 hours under favorable conditions, but only 4 generations when the temperatures are low.

Even this rate of multiplication, which might apply for all autotrophic bacteria, is much higher than that of unicellular green protists. The speed of transmission of geochemical energy should be correspondingly greater, and we should therefore expect that the bacterial mass in the biosphere would far exceed the mass of green eukaryotes. In addition, the phenomenon observed in the seas (§51) should occur, with bacteria predominating over green protista, in the same way that unicellular algae predominate over green metaphytes.

98 In fact, however, this is not what actually happens. Restraints are imposed on the accumulation of this form of living matter for reasons similar to those that allow green metaphytes to predominate over green protista on land.

Monera are ubiquitous, existing throughout the ocean to depths far beyond the penetration of solar radiation, and they are diverse enough to include nitrogen, sulfur, and iron bacteria, which are not as common as other types of bacteria. One is led to conclude that bacterial abundance is a ubiquitous and constant feature of the Earth's surface.

We find confirmation of this in the very special conditions of nutrition on which bacterial existence depends. Chemoautotrophic bacteria receive all the energy needed for life by completing the oxidation of unoxidized and partially-oxidized natural compounds of nitrogen, sulfur, iron, manganese, and carbon. Oxygen-depleted matter (vadose minerals of these elements) needed by these bacteria can never be amassed in sufficient quantities in the biosphere, because *the domain of the biosphere is a region of oxidation*, saturated by free oxygen created by green organisms. In this medium, the most thoroughly oxidized compounds are the most stable forms. Autotrophic microorganisms are consequently forced to actively search for a favorable medium, and their various adaptations result from this requirement.

While bacteria can secure the energy needed for life by transforming partially-oxidized into completely-oxidized matter, the number of chemical elements in the biosphere capable of such reactions is limited. Stable final forms of oxygen-rich compounds are also created independently of bacteria by purely chemical processes, since the biosphere is intrinsically a medium in which such molecular structures are stable.

99 *Autotrophic bacteria are in a continual state of deprivation.* This results in numerous adaptations of life, as demonstrated in aqueous basins, mineral-water springs, sea water, and damp soil, where we observe curious secondary equilibria between sulfate-reducing bacteria and autotrophic organisms that oxidize sulfides. The former establish the conditions of existence for the latter. Numerous cases of such secondary equilibria show that this phenomenon is a part of an orderly mechanism. Living matter has developed these equilibria as a result of the immense vital pressure of autotrophic bacteria seeking a sufficient quantity of the ready-made, oxygen-poor compounds (§29). Living matter itself has created these compounds within an inert medium in which such compounds were not originally present.

In the ocean, an identical exchange occurs between autotrophic bacteria that oxidize nitrogen and heterotrophic organisms that deoxidize the nitrates. This forms one of the marvelous equilibria of the hydrosphere's chemistry.

The ubiquity of these organisms shows their immense geochemical energy and speed of vital transmission; their limited distribution indicates the shortage of oxygen-poor compounds in the biosphere, where green plants are continuously releasing excess free oxygen. If these organisms do not constitute a considerable mass of living matter, it is only because it is physically impossible for them to do so: the biosphere lacks sufficient quantities of the compounds needed for their existence.

Although the precise relationships still elude us, there must be definite quantitative relationships between the amount of matter in the biosphere composed of photoautotrophs and that composed of autotrophic bacteria.

100 The opinion is often expressed that these curious organisms are representatives of the oldest organisms, and having an even earlier origin than green plants. One of the most eminent naturalists and thinkers of our time, the American H. F. Osborn,[223] has recently reiterated such ideas.

The role of autotrophic bacteria in the biosphere contradicts this view, however. The strict connection between their presence or absence and the presence of free oxygen proves their dependence on green organisms and on the energy of solar radiation. This dependence is equally strong for fungi and heterotrophic bacteria, and for animals which feed upon the materials produced by green plants.

The character of their functions in the general economy of nature shows also their derivative importance, in comparison to green plants. Their importance is enormous in the biogeochemical history of sulfur and nitrogen, the two indispensable elements for construction of the essential matter of the protoplasm — amino acids and proteins. If the activity of these autotrophic organisms were to stop, life would, perhaps, be quantitatively reduced; but it would remain a powerful mechanism in the biosphere, because the necessary vadose compounds (nitrates,[224] sulfates, ammonia, and hydrogen sulfide) are created in great quantities independently of life.

Without wishing to anticipate the question of autotrophy (§94) and the beginnings of life on Earth, it is very probable that autotrophic bacteria depend upon photosynthetic organisms, and that their origin was derivative from theirs.[225] Everything indicates that these autotrophic organisms are vital forms which complete the utilization of the energy of solar radiation by perfecting the "solar radiation-green organism" mechanism, and are not a form of life that is independent of radiant energy from the cosmos.

The whole heterotrophic world, with its innumerable forms of animals and fungi, millions of species, is analogous to the same process.

101 This fact is also clearly demonstrated by the nature of distribution of living matter in the biosphere.

This distribution is determined entirely by the field of stability of the green vegetation; in other words, by the region of the planet that is saturated with solar radiation. The principal mass of living matter is concentrated in this region, including heterotrophic organisms and autotrophic bacteria, the existence of which is closely connected to the free oxygen or the organic compounds created by green organisms.

223 See Osborn, 1917.

224 See Canter, 1996.

225 Another facet of the 'eternal life' assumption.

Heterotrophic organisms and autotrophic bacteria are capable of penetrating regions of the biosphere where solar radiation and green life are not present. A great number of such organisms inhabit only these dark regions of the biosphere. It is commonly assumed, probably correctly, that they have migrated to these areas from the sunlit surface, gradually adapting themselves to the new conditions of life. Morphological studies of deep sea and cave-dwelling animals indicate (sometimes irrefutably) that such fauna are derived from ancestors that formerly inhabited the illuminated regions of the planet.

From the geochemical point of view concentrations of life that do not contain green organisms have particular importance.[226] Some of these concentrations make up the *benthic film* at the base of the hydrosphere (§130), the lower parts of the *littoral concentrations of the oceans*, and the living films at the bottom of the continental aqueous basins (§158). We shall see the immense role in the chemical history of the planet that these concentrations play. We can be certain, however, that they are closely connected, directly or indirectly, with organisms of the green regions. The morphology and paleontology of this second-order living matter suggest its derivation from organisms inhabiting the sunlit regions of the planet, as mentioned above.

There is also another way in which solar radiation is the basis of its daily life. These deep vital films have a close relationship to the organic debris that falls from the upper portions of the ocean, and reaches the bottom before it has had time to decompose. Anaerobic organisms in the bottom film depend upon this debris for food. The parts of the planet illuminated by the sun are thus the primary energy source for these films.

Free oxygen from the atmosphere also penetrates to the bottom of the sea. Everything indicates that these phenomena of the benthic regions are in a state of perpetual evolution, and that their field of influence is becoming more and more vast. A slow and continuous movement of living matter, from the green bed into various azoic[227] regions of the planet, seems to have taken place throughout geologic time; and at the present stage, the domain of life is being extended by benthic organisms.

102 The creation of new forms of luminous energy by heterotrophic living matter may be one manifestation of this process. The phosphorescence of organisms (bioluminescence) consists of wavelengths which overlap those of solar radiation on the Earth's surface. This secondary luminous radiation

226 Such communities are indeed widespread; for example sea floor sulfur bacteria mats (Fossing, et. al., 1995) and the now famous vent biotas (Gould and Gould, 1989).

227 There are very few surface regions of the planet that are truly azoic. As noted by Thomas Gold (1996), a strong argument against sending a manned mission to Mars is that Earth bacteria might escape from the astronauts and render it impossible for us to tell if Mars has or ever had any endemic life forms: "Mars as a 'Rosetta stone' for the origin of life would be lost forever."

allows green plankton over an area of hundreds of square kilometers to produce their chemical work during times when solar energy does not reach them. The phosphorescence becomes more intense with depth.

Is the phosphorescence of deep-sea organisms a new example of the same mechanism? Is this mechanism causing photosynthetic life to revive, several kilometers below the surface, by transmitting solar energy to depths it could not otherwise reach? We do not know, but we must not forget that oceanographic expeditions have found green organisms living at depths far beyond the penetration of solar radiation. The ship "Valdivia" found, for example, *Halionella* algae living at a depth of about two kilometers.[228]

The transportation by living matter of luminous energy into new regions, in the form of thermodynamically unstable chemical compounds, and also of secondary luminous energy — phosphorescence — causes a slight provisional extension of the domain of photosynthesis. This is analogous to the luminous energy created by human civilization, which is used by green living matter, but has not yet significantly affected photosynthesis on the planet.[229]

Now we shall turn from the discussion of green living matter, in order to deal with the rest of the living world.

The Limits of Life

103 The field of stability of life extends beyond the limits of the biosphere, and the independent variables which determine the stability (temperature, chemical composition, etc.) attain values well beyond the characteristic biospheric extremes of these quantities.

The field of stability of life is the region in which life can attain its fullest expansion. This field seems to be neither rigorously determined nor constant.

The ability of organisms to adapt to previously lethal conditions after a number of generations is a characteristic of living matter.[230] We cannot study this subject experimentally, because millions of years are required for the process of adaptation.[231] Living matter, in contrast to inert matter, displays a mobile equilibrium. This equilibrium, acting over stretches of geologic time, exerts a pressure upon the surrounding environment.

In addition, the field of stability of life is clearly divided into the field of gravity for the more voluminous organisms, and the field of molecular force for the smaller organisms such as

228 This is a remarkable observation and would strain belief if not for recent proposals from western scientists that photosynthesis *originated* at deep sea vents at vanishingly low levels of ambient light (Dover, 1996; Zimmer, 1996). The shrimp *Rimicaris exoculata* ("eyeless fissure shrimp") has unusual, large, strip-shaped eyes along its back (not in the ordinary position of eye sockets) and it can actually see light (invisible to humans) emanating from deep sea hydrothermal vents. These arthropods apparently use the light to keep a safe distance from the superheated vent emissions. The most primitive forms of bacterial chlorophyll absorb the most light at the frequencey bands measured at hydrothermal vents, leading Dover to propose that photosynthesis evolved at such vents. Vernadsky might have frowned on such a western-style, extrapolationistic inference.

229 Flourishing communities of photosynthesizing mosses are known which use as an energy source the lights installed for tourists in Howe Caverns in upstate New York (–M. McMenamin).

230 Unfortunately, this sentiment was carried to an extreme by T. Lysenko (1948) in his unsuccessful attempts to "force" crops to adapt to harsh climates. Vernadsky played no part in this fiasco.

231 See Case, 1997.

microbes and ultramicrobes (on the order of 10^{-4} mm long). The lives and movements of the latter are primarily determined by luminous and other radiations. Although the sizes of these two fields are not well documented, we know that they must be determined by the tolerances of organisms.

We shall consider the most important characteristics of both fields of stability (that is, the field of stability of life and the field of stability of the biosphere): 1. temperature; 2. pressure; 3. state of matter of the medium; 4. the chemistry of the medium; and 5. luminous energy.

104 We must now distinguish two kinds of conditions: 1. those which do not exceed life's capacity to endure and to exercise its functions (i.e., conditions which cause suffering but not death); and 2. conditions which allow life to multiply.

Due to the genetic links uniting all living matter, these conditions may be about the same for all organisms; the field is, however, much more limited for green vegetation than for heterotrophic organisms. The limit is actually determined by the physico-chemical properties of the compounds constituting the organism, and specifically by their stability. In some cases, it appears that the mechanisms formed by compounds that control the functions of life[232] are first to be destroyed under difficult conditions. Both the compounds and the mechanisms they catalyze have been ceaselessly modified by adaptation during geological time.

Some extreme examples of the survival of particular organisms will illustrate the maximum vital field as it is presently known.

105 Certain heterotrophic organisms (especially those in the latent state, such as fungal spores) demonstrate the highest temperature tolerance, and can withstand about 140° C. This limit varies with the humidity of the habitat.

L. Pasteur's experiments[233] on spontaneous generation showed that some microbial spores were not destroyed by raising the temperature to 120° C in a humid medium. Destruction required no less than 180° C (Ducloux[234]).* In Christensen's experiments, soil bacteria resisted temperatures of 130° C for five minutes, and 140° C for one minute.[235] The bacterial spores described by Zettnow[236] were not destroyed after exposure to steam flow for twenty-four hours (V. L. Omeliansky[237]).

At low temperatures, the field of stability is greater. Experiments at the Jenner Institute in London showed that bacterial spores

232 Such as enzymes.

233 For descriptions of these experiments, see Vallery-Rodot, 1912; Compton, 1932; Pasteur 1876.

234 See Ducloux, 1905, 1908.

* This impression of Pasteur's collaborators at the time of his celebrated discussion with G. Fouche seems to have greater importance for the determination of the maximum temperature in the vital thermal field, than in experiments on pure cultures. It is based on the study of the properties of hay infusions, which are closer to the complex medium of life on Earth's crust than are pure cultures.

235 See Christensen and Larsen, 1911; and Christensen, 1915.

236 See Zettnow, 1912.

237 See Omeliansky, 1923.

remained stable for twenty hours in liquid hydrogen (-252° C). A. MacFadyen reports microorganisms which remained intact for many months in liquid air (-200° C).[238] In P. Becquerel's experiments, *fungal spores* remained in vacuum at -253° C for seventy-two hours without losing their viability.[239] Similarly, the most diverse plant seeds have survived ten-and-a-half hours in vacuum at an even lower temperature . . . -269° C.

We can thus estimate a range of 450°C[240] as the thermal field within which certain vital forms can survive. The range is clearly less for green vegetation. There are no precise experiments on this subject, but it is doubtful that the range is greater than 150-160° C (from -60° to +80°).

106 The limits of pressure in the vital field are large. In experiments by G. Khlopin and G. Tammann[241], fungi, yeast, and bacteria withstood a pressure of 3000 atmospheres without apparent change in properties. Yeast survived 8000 atmospheres. At the other extreme, seeds and spores can be preserved for long intervals in a vacuum. There seems to be no difference between heterotrophic and photoautotrophic organisms in this regard.

107 The importance of certain radiant energy wavelengths for plants has often been pointed out. This is true for the entire biosphere. Photosynthesizers perish relatively quickly when this radiation is absent. Heterotrophic organisms, and at least certain autotrophic bacteria, can live in the dark; but the character of this darkness (long-wavelength, infrared radiation) has not been studied. On the other hand, we know that short and highly energetic wavelengths of electromagnetic radiation are an insurmountable barrier to life.

The medium characterized by very short, ultraviolet wavelengths is inanimate (§114). Becquerel's experiments have demonstrated that this radiation kills all forms of life in a very short time.[242] Interplanetary space, where these rays are present, is inaccessible to all forms of life adapted to the biosphere, although neither the temperature, pressure, nor chemical character of this space presents any obstacle.[243] The confines of life in the various regions of radiant energy must be studied in detail, as indicated by what we know about the relationship between life in the biosphere and solar radiation.

108 The scale of chemical changes that life can undergo is enormous. The discovery by L. Pasteur of anaerobic organisms proved

238 See MacFadyen, 1902.

239 See Becquerel, 1910.

240 The figures in this paragraph are from the French edition of 1925. See Van Hise, 1904, p. 52; and Hann, 1883.

241 See Tamman and Khlopin, 1903.

242 Papers by Daly and Minton (1996) show that some organisms can indeed withstand hard penetrating radiation.

243 Vernadsky was thus willing to entertain the concept of panspermia.

that life can exist in a medium without free oxygen. The assumed limits of life were greatly enlarged by Pasteur's discovery.[244]

The autotrophic organisms discovered by S. Vinogradsky showed that life could exist in a purely mineral medium, containing no preexisting organic compounds.[245]

Spores and seeds seem to be able to remain perfectly intact, as latent vital forms, for an indefinite time in a medium devoid of gas or water.

It is also possible for various forms of life to live with impunity in the most diverse chemical media. *Bacillus boracicolla,* in the hot boracic-water springs of Tuscany, can live in a saturated solution of boric acid, and also easily endures a ten percent solution of sulfuric acid at environmental temperatures.[246] There are many organisms, chiefly fungal molds, which live in strong solutions of salts fatal to other organisms, such as saturated solutions of the sulfate, nitrate, and niobate of potassium.

Bacillus boracicolla, mentioned above, resists solutions of 0.3% mercury chloride; other bacteria and protists survive in saturated solutions of mercury chloride;[247] yeast can live in a solution of sodium hydrofluorate. The larvae of certain flies survive 10% solutions of formalin, and there are bacteria which multiply in an atmosphere of pure oxygen. These phenomena are poorly known, but they demonstrate the apparently unlimited adaptability of living forms.

We are only speaking here of heterotrophic organisms. The development of green organisms demands the presence of free oxygen (sometimes in aqueous solution). Saturated saline solutions make the development of this form of life impossible.

109 Although certain forms of life in a latent state can survive in an absolutely dry medium, water in liquid or vapor form is essential for growth and multiplication. In addition to the obvious fact that green life cannot exist without water, it can be noted that the geochemical energy of organisms, as shown by multiplication, changes from a potential to an active form only in the presence of water containing gases needed for breathing.

The foundation of all life, photosynthetic green life, cannot exist without water. The mechanism of the action of water has recently become clearer, through an understanding of the acid-base balance and the degree of ionization of aqueous solutions.[248]

The role of these phenomena is enormous, because most of the mass of living matter is concentrated in the natural waters of

[244] See Pasteur, 1876.

[245] See Vinogradsky, 1888a, 1888b, and 1989..

[246] See Bargagli Petrucci, 1914.

[247] See Besredka, 1925.

[248] In recent years the important role of the structure of water has also been clarified, depending upon the condition of the geomagnetic pole and solar activity. –A. I. Perelman.

the biosphere, and the living conditions of all organisms are closely related to natural aqueous solutions. The matter of organisms is formed principally of aqueous salt solutions.* Protoplasm may be considered an aqueous suspension in which coagulations and colloidal changes occur. The phenomena of ionization are ubiquitous. Because of continuous, reciprocal action between the internal liquids of the organisms and the surrounding aqueous solutions, the relative ionizations of the two media are of great importance. Subtle methods of recording exact changes in ionization provide an excellent way to study the principal medium of life.

Sea water contains about 10^{-9}% (H+); it is thus slightly alkaline, and is continuously maintained so, in the presence of numerous ongoing chemical processes.

This degree of ionization is very favorable for the life of marine organisms; the slightest variations always have repercussions in living nature, positive or negative, depending on the organism.

It is clear that life can exist only within limits of ionization between 10^{-6} and 10^{-10}% (H+).[249]

110 The physical state of the medium is extremely important for life processes. Life in latent form can be preserved in solid, liquid, or gaseous states, and in a vacuum. Seeds can be preserved for a certain time, without gaseous exchange, in all states of matter. But the fully-functioning living organism requires gaseous exchange (respiration), and stable conditions for the colloidal systems that form its body. In solid media, living organisms are found only in porous bodies where there is access to gaseous exchange. Due to the very small dimensions of many organisms, fairly compact media can be inhabited; but a liquid solution or colloid cannot maintain life if it lacks gas.

We once again encounter the exceptional importance of the gaseous state of matter, a point frequently made in this essay.

The Limits of Life in the Biosphere

111 Thus far, we have seen that the biosphere, by structure, composition, and physical makeup, is completely enclosed by the domain of life, which has so adapted itself to biospheric conditions that there is no place in which it is unable to manifest itself in one way or another.

This statement does not hold true under temporary, abnormal circumstances, such as would prevail during times of erupting

* Organisms contain 60 to 99% water by weight, and thus are composed from 80-100% of aqueous solutions or suspensions.

249 Contemporary data indicate that these limits are in fact much broader.

volcanoes and lava flows. Toxic volcanic exhalations (hydrochloric and hydrofluoric acids, for example), and hot springs which accompany volcanic action, are examples of such temporary phenomena; the absence of life caused by them is also temporary. Analogous phenomena of longer duration, such as permanent thermal sources with temperatures of about 90° C, are inhabited by organisms adapted to these conditions.

Natural saline solutions with concentrations more than 5% may not be permanently inanimate; we simply do not know whether they are or not. The Dead Sea in Palestine is regarded as the largest saltwater basin of its kind. There is proof, however, that certain naturally acidic waters (containing hydrochloric and sulfuric acid), whose ionization is at least 10^{-11}% (H+), must be inanimate[250] (§109). The extent of such dead zone is, however, insignificant when compared to the planet as a whole.

112 The terrestrial envelope occupied by living matter, which can be regarded as the entire field of existence of life, is a continuous envelope, and should be differentiated from discontinuous envelopes such as the hydrosphere.

The field vital stability is, of course, far from completely occupied by living matter; we can see that a slow penetration of life into new regions has occurred during geological time.

Two regions of the field of vital stability must be distinguished: 1. the region of temporary penetration, where organisms are not subject to sudden annihilation; and 2. the region of stable existence of life, where multiplication can occur.

The extreme limits of life in the biosphere probably represent absolute conditions for all organisms. These limits are reached when any one of these conditions, which can be expressed as independent variables of equilibrium, becomes insurmountable for living matter; it might be temperature, chemical composition, ionization of the medium, or the wavelength of radiations.

Definitions of this kind are not absolute, since adaptation gives organisms immense ability to protect themselves against harmful environmental conditions. The limits of adaptation are unknown, but are increasing with time on a planetary scale.

Establishing such limits on the basis of known adaptations of life requires guesswork, always a hazardous and uncertain undertaking. Man, in particular, being endowed with understanding and the ability to direct his will, can reach places that are inaccessible to any other living organisms.[251]

Given the indissoluble unity of all living beings, an insight

250 This is not correct, although the modern correction only further validates Vernadsky's basic point. The range indicated (pH = 5-9) is the most favorable for life, but living organisms can exist both in more-acid (bacteria, down to pH = 1 or less) and in more-alkaline media. The Dead Sea is now known to be inhabited by a variety of peculiar bacteria.

251 In Antarctica, at the Soviet "Vostok" station, people worked during the winter without special adaptation in conditions of oxygen starvation, with winds 8-10 m/sec (18-22 m.p.h.) and temperatures -80° C. The equivalent still air temperature is -130° C. –A. I. Perelman.

flashes upon us. When we view life as a planetary phenomenon, this capacity of *Homo sapiens* cannot be regarded as accidental.[252] It follows that the question of unchanging limits of life in the biosphere must be treated with caution.

113 The boundaries of life,[253] based upon the range of existence of contemporary organisms and their powers of adaptation, clearly show that the biosphere is a terrestrial *envelope*. For the conditions that make life impossible occur simultaneously over the whole planet. It is therefore sufficient to determine only the upper and lower limits of the vital field.

The upper limit is determined by the radiant energy which eliminates life. The lower limit is formed by temperatures so high that life becomes impossible. Between these limits, life embraces (though not completely) a single thermodynamic envelope, three chemical envelopes, and three envelopes of states of matter (§88). The above limits clearly reveal the importance of the last three, the troposphere, hydrosphere, and upper lithosphere. We will take these as the basis of the exposition to follow.

114 By all appearances the natural forms of life cannot pass beyond the upper stratosphere. As Table I shows (§88), there is a paragenetic envelope above the stratosphere in which the existence of chemical molecules or compounds is unlikely. This is the region of the maximum rarefaction of matter, even if we accept Prof. V. G. Fesenkov's calculations,[254] which suggest that it contains greater quantities of matter than formerly supposed (1923-1924). He states that there is one ton of matter per cubic kilometer at a height of 150 to 200 kilometers.* The new mode of occurrence of chemical elements in these regions results not only from dispersion, reduction of collisions, or lengthened free trajectories of gaseous particles, but also from the powerful action of ultraviolet and other solar radiation and also, perhaps, from the activity of cosmic radiation (§8). Ultraviolet radiation is a very active chemical agent, which in the 160-180 nm range destroys all life, even spores that are stable in dry media or a vacuum. It seems certain that this radiation penetrates the upper stratosphere.

115 This radiation reaches no further down because of its complete absorption *by ozone*, which is continually formed in relatively large quantities from free oxygen (and perhaps water) by the action of this same ultraviolet radiation.

252 We see here again Vernadsky's challenge to chance-driven causality; "chance does not exist" as he put it earlier. The idea of arogenesis or progressive evolution has a long history in Russian evolutionary biology (see comment by Liya N. Khakhina in "Note on translation and transliteration" by Mark McMenamin and Lynn Margulis, p. xxix in L. N. Khakhina, 1992). This idea of progressive evolution, most fully articulated in the west by P. Teilhard de Chardin in his use of the term *complexification*, is anathema to some contemporary western evolutionary biologists. The question remains open, especially in the minds of Russians such as Khakhina.

253 Vernadsky devoted an entire article to this subject: Vernadsky, 1937.

254 See Fesenkov, 1976.

* According to other calculations, the figures are more than a thousand times less — one ton per one hundred cubic kilometers; one kilogram per 200 cubic kilometers.

According to Fabry and Buisson, the entire amount of ozone in a pure state would form a layer five millimeters thick.[255] Nevertheless, this small amount, dispersed among the atmospheric gases, is enough to halt these fatal radiations.

The ozone is recreated as fast as it is destroyed, because the ultraviolet radiation meets an abundance of oxygen atoms lower in the stratosphere. Life is thus protected by an *ozone screen* five millimeters thick,[256] which marks the natural upper limit of the biosphere.

The free oxygen necessary for the creation of ozone is formed in the biosphere solely through biochemical processes, and would disappear[257] if life were to stop.[258] *Life creates both the free oxygen in the Earth's crust, and also the ozone that protects the biosphere from the harmful short-wavelength radiation of celestial bodies.*

Obviously, life's latest manifestation — civilized man — can protect himself in other ways, and thus penetrate beyond the ozone screen with impunity.

116 The ozone screen determines only the potential upper limit of life, which actually stops well below this atmospheric limit.

Green autotrophics do not develop above the forests, fields, prairies, and grasses of land. There are no unicellular green organisms in the aerial medium. The oceans throw green plankton a short distance upward, but only accidentally. Organisms can raise themselves higher than green vegetation only by mechanical means created for flight. Even in this way, green organisms cannot penetrate the atmosphere for any great distance or length of time.[259] For example, the spores of the conifers and cryptogams are probably the largest mass of green organisms dispersed and lifted up by the winds. Sometimes they reach great heights, but only for short periods. These are the smallest spores, and contain little or no chlorophyll.

The green layer is the upper limit of transformation of solar radiation, and is situated on the surface of the land and in upper layers of the ocean. The range of this layer has become more vast over geologic time, but it does not rise into the atmosphere to any great extent.

Due to the proclivity of the green autotrophs to maximize the capture of solar energy, they have penetrated very efficiently into the lower layers of the troposphere, rising to a height of over 100 meters in the form of large trees and stands of vegetation 50

255 See Fabry and Buisson, 1913.

256 Contemporary observations have demonstrated a decrease in the ozone layer of three to four percent during the interval 1969–1993. The largest reduction in the ozone shield occurs over the Antarctic, where there is an ozone hole 40 x 10^6 square kilometers in area (Karol', Kiselev and Frol'kis, 1995).

257 A certain amount of free oxygen is formed in the upper layers of the atmosphere by the action of cosmic rays, possible reactions being $N + p \longrightarrow O + h\upsilon$; $O \longrightarrow N + h\upsilon$; and $N + p \longrightarrow O + h\upsilon$ (p = proton; $h\upsilon$ = types of radiation). While the magnitude of this process cannot yet be calculated, it obviously plays only an insignificant role in the planetary oxygen balance, since the amount of oxygen in the atmosphere agrees within an order of magnitude with that arising from green living matter. –A. I. Perelman.

258 In 1856 C. Koene hypothesized that atmospheric oxygen was the result of photosynthesis. Vernadsky gave this idea special attention, and from the perspective of geochemistry (Voitkevich, Miroshnikov, Povarennykh and Prokhorov, 1970). Vernadsky was the first to demonstrate the biogenic origin of atmospheric oxygen (Vernadsky, 1935; Oparin, 1957, p. 157).

259 Study of life in the atmosphere has been continued by S. V. Lysenko (1979).

or more meters in height. These vital forms have been in existence since the Paleozoic Era.[260]

117 The principal mass of living matter that penetrates the atmosphere belongs to the second order, and includes all flying organisms. For millions of years life has penetrated the atmosphere principally in the form of the very small bacteria, spores and flying animals. Larger concentrations, mostly in the latent state as spores, can only be observed in regions penetrated by dust. According to A. Klossovsky,[261] dust reaches an average height of five kilometers; O. Mengel[262] states that this distance is 2.8 km. In any case, it is chiefly inert matter.

Air on the tops of mountains is very poor in organisms, but some do exist there. L. Pasteur found only an average of four to five pathogenic microbes per cubic meter,[263] when such air was cultured. Fleming has found, at the most, only one pathogenic microbe per three liters at a height of four kilometers[264].

It seems that the microflora of the upper layers are relatively poor in bacteria, and rich in yeast and fungi[265]. Microflora certainly penetrate beyond the average boundary of the dusty atmosphere (5 km), but there are few precise observations. This flora might be carried to the limits of the troposphere (9-13 km) since the movement of air currents reaches this height. It is unlikely that the ascent of this material has played any role whatsoever in the Earth's history, in view of the latent state of most of these organisms, and the barely detectable numbers of them present.

118 It is not clear whether animals go beyond the troposphere, though they sometimes reach its upper regions above the highest mountain summits.

According to A. Humboldt, the condor flies to heights of seven kilometers. He has also observed flies on the summit of Mt. Chimborazo (5882 meters).[266]

The observations on avian flight by Humboldt and others have been challenged by modern studies of bird migrations. The latest observations of A. F. R. Wollaston and members of the British Mt. Everest expedition, however, prove that certain alpine birds of prey soar around the highest summits (7540 m). Himalayan crows were seen up to 8200 meters.[267]

These, however, are values for particular species. Most birds, even in mountainous country, go no higher than 5 km. Aviators have not found them above 3 km; eagles have been observed this high.[268]

260 The first forests appear in the Late Devonian. (Snigirevskaya, 1988; McMenamin and McMenamin, 1993.)

261 See Klossovsky, 1908, 1914.

262 See Mengel, 1923.

263 See Pasteur, 1876. Here there is a discordance between the Russian (1926) and the French editions of *The Biosphere*. In the Russian it reads "per cubic meter," in the French "per cubic centimeter." The former is apparently the correct version.

264 See Fleming, 1908.

265 See Omeliansky, 1923.

266 See Humbolt, 1859.

267 See Howard-Bury, 1922.

268 A griffon vulture is known to have collided with an aircraft at an altitude of 12.5 kilometers (A. Lapo, written communication).

Butterflies have been seen at 6400m; spiders at 6700 m; the green fly, at 8200 m. Certain plants (*Arenaria muscosa* and *Delphinium glacial*) live at 6200 to 6300 m (Hingston,[269] 1925).[270]

119 It is man who ascends to the greatest heights of the stratosphere, unwittingly taking with him the forms of life that accompany his body or its products.

The region accessible to man is enlarging with the development of aerial navigation, and now extends beyond the ozone screen.

Sounding balloons, which have reached the greatest heights, always contain some representatives of life. A balloon of this kind launched at Pavia, December 17, 1913, went to 37,700 m.

Man, himself, has risen above the highest mountain. In 1875, G. Tissandier,[271] and 1868, J. Glaisher, almost achieved this in aerostatic balloons (8600 m and 8830 m, respectively[272]); airplanes have reached the limits of the troposphere. The Frenchman Callisot, and the American J. A. Macready in 1925, rose to 12,000 and 12,100 m; and this record will obviously soon be surpassed.[273]

With respect to permanent human collections, villages reach 5100 and 5200 m in Peru and Tibet; railways, 4770 m in Peru. Oat fields exist at 4650 m.

120 In short, life in the biosphere reaches its terrestrial boundary, the ozone screen, only in rare circumstances. Both the stratosphere and the upper layers of the troposphere are essentially inanimate.

No organism lives continuously in the air. Only a thin layer of the air, usually well below 100 meters thick, can be considered to contain life.

The conquest of the air is a new phenomenon in the geological history of the planet. It could not have happened without the development of subaerial terrestrial organisms — plants (Precambrian origin?), insects, flying vertebrates[274] (Paleozoic origin?), and birds (Mesozoic origin). Evidence for the mechanical transport of microflora and spores dates from the most distant geological periods. But only when civilized humanity appeared did living matter make a great stride toward conquest of the entire atmosphere.

The atmosphere is not an independent region of life. From the biological point of view, its thin layers are part of the adjacent hydrosphere and lithosphere. Only in the upper part of the

269 Vernadsky's exact reference could not be located but see Hingston (1925).

270 In mountains above 6000 m other animals are also found, although large animals enter this zone only while migrating from one slope to another. Some species of mammals (e.g. pikas), birds, and insects inhabit regions above 6000 m during the summer, in areas free of ice and snow. –A. I. Perelman.

271 See Tissandier, 1887, 1887-1890. Gaston Tissandier was a chemist. He and his brother Albert, with three others, remained in the air for 23 hours in 1875. In 1824 G. Tissandier rose to a height of 9150 meters. He was the sole survivor of this ascent, the other two with him dying of asphyxia.

272 This ascent was made without breathing apparatus. The height has been disputed.

273 As indeed it was, culminating of course in America's NASA Apollo space program that landed men on the moon in the late 1960's and early 1970's. Both Callisot and Macready were early aviators.

274 See Padian, 1985.

lithosphere do atmospheric layers take part in concentrations of life and living films. (§150)

The enormous influence exercised on the history of the atmosphere by living matter is not related to the immediate presence of life in the gaseous medium, but to gaseous exchange — the biological creation of new gases, and their liberation and absorption in the atmosphere. This occurs both in the gaseous layer adjacent to the Earth's surface, and in gases dissolved in natural waters.

The final, grand effect — the occupation of the entire gaseous envelope of the planet by vital energy, resulting from diffusion of oxygen and other gases produced by life — is a result of the gaseous state of matter, and not of the properties of living matter.

121 Theoretically, the lower limit of life on Earth should be just as distinct as the upper limit set by the ozone screen, and should be determined by temperatures that make the existence and development of an organism impossible.

The temperature of 100° C that clearly marks this barrier is usually found 3 to 3.5 km below the Earth's surface, though at certain places it occurs at about 2.5 km. We may assume that living creatures in their present forms cannot exist at a depth greater than 3 km below[275] the surface of the earth.

The level of this 100° C planetary boundary is deeper under the ocean, which has a mean depth of 3.8 km, and bottom temperatures a few degrees above zero. The temperature barrier to life under ocean regions would therefore occur at depths of 6.5 to 6.7 km, assuming thermal conditions to be the same as under land areas. Actually, the rise in temperature with depth seems to take place more rapidly under the sea,[276] and it is unlikely that layers accessible to life extend more than 6 km below the surface of the hydrosphere.

This limit of 100° C is conventional. Some organisms on the Earth's surface are able to multiply at temperatures of 70° to 80°, but none are known to have adapted to permanent existence at 100° C.[277]

Thus it is unlikely that the lower limit of the biosphere exceeds a depth of 2.5 to 2.7 km on land, and 5 to 5.5 km in the oceans.

This limit is probably determined by the temperature,[278] rather than by oxygen deficiency in the deep regions, because absence of oxygen is no obstacle to life. Free oxygen vanishes at shallow depths beneath the continents, and is rarely observed even a few hundred meters beneath the surface. It is certain,

275 Recent studies have pushed close to this limit, with live bacteria occuring at depths up to 2.8 kilometers beneath the surface of Earth (Kerr 1997; Anonymous, 1996b; and Fredricson and Onstatt, 1996). Acetogenic bacteria occur in sediments to depths of 888 meters (See Chapelle and Bradley, 1996; Monastersky 1997).

276 Vernadsky means here that the geo-thermal gradient is steeper in oceanic crust than in continental crust.

277 This still appears to be the case, even for hyperthermophilic bacteria (which can survive exposure to 110°C; Peak et al. 1995).

278 And also by salt concentrations (Lapo, 1987). For contemporary discussion of the deep hot biosphere, see Delaney et al., 1994; Pedersen, 1993; and Gold and Soter, 1980.

however, that anaerobic life penetrates to much greater depths. The independent researches of E. S. Bastin[279] of the United States and N. Ushinsky of Russia[280] (1926-1927) have confirmed the earlier observations of F. Stapff[281] that anaerobic flora exist more than a kilometer below the Earth's surface.

122 High temperatures are an insurmountable, though theoretical, limit for the biosphere. Other factors, taken together, have a much more powerful influence on the distribution of life, and prevent it from reaching regions which would otherwise be accessible to it from a thermal standpoint.

Certain macroscopic organisms have created a curious geologic phenomenon as they have penetrated into the depths. These dark regions of the planet have been populated by specific organisms which are geologically young, and the tendency to reach downward continues.

In a manner analogous to the situation at the upper limit of the biosphere, life is descending slowly but ineluctably to greater depths. The lower limit it is approaching is, however, further away than the upper limit.

Green organisms obviously cannot leave the illuminated regions of the Earth's surface. Only heterotrophic organisms and autotrophic bacteria can go lower.

Life penetrates the depths of the earth in a different manner than in the ocean. Dispersed animal life exists at very great oceanic depths, depending upon the shape of the sea floor. A sea urchin[282] — *Hyphalaster perfectus* — has been found at a depth of 6035 meters. Benthic aquatic forms can penetrate into the deepest trenches, but so far living organisms have not been found below 6500 meters.*[283] The entire ocean contains dispersed bacteria, which at depths more than 5500 meters are concentrated on the ocean bed. Their presence in the mud of the deepest trenches has not been proven definitely, however.[284]

123 Life does not go so deep on land, primarily because free oxygen does not penetrate very far. Dissolved free oxygen in the ocean (where its proportion, relative to nitrogen, is greater than in the atmosphere) is in direct contact with the outside air. Atmospheric oxygen reaches the deepest ocean trenches (10 km), and losses are continually replaced from the atmosphere by solution and diffusion. The layer of penetration of free oxygen is the thin upper layer of seafloor mud. (§141)

279 See Bastin, 1926.

280 See Ushinsky, 1926; Ginzburg-Karagicheva, 1926; Bogachev, 1927; Ginzburg-Karagicheva, 1927; and Ushinsky, 1927.

281 See Stapff, 1891.

282 The Soviet exploration vessel "Vityaz" found abundant organisms on the bottom of deep trenches. The American submersibles have confirmed these results.

* The depth of the ocean reaches nearly
10 km: a depth of 9.95 km has recently been recorded near the Kurile Islands.
The greatest previously-known depth was 9.79 km near the Philippines.
[A. I. Perelman gives the following data with respect to the deepest parts
of the world oceans: Mariana trench, 11,022 m; Tonga trench, 10,882; Kurile-Kamchatka trench, 10,542; Philippine trench, 10,497.]

283 Expeditions of the *Vityaz* (USSR) and the *Galatea* (Denmark) in the 1940s and 1950s demonstrated the presence of life in the greatest oceanic depths, i. e., at depths up to 11,022 meters.

284 It has been demonstrated more recently (Zaninetti, 1978).

Free oxygen disappears very quickly with depth on land, being rapidly absorbed by organisms and oxygen-hungry organic and other reduced compounds. Water from sources one or two kilometers deep lacks free oxygen. A sharp division is observed between vadose water that contains free oxygen from the air, and phreatic water that contains none, but this has not yet been fully clarified.[*][285] Free oxygen usually penetrates the entire soil and part of the subsoil. Its level is nearer the surface in marshy and swampy soils. According to Hesselman, such soils in our latitudes lack free oxygen below a depth of 30 cm.[286] Oxygen has been found in subsoil at a depth of several meters (sometimes more than 10 m) where there is no obstacle to its passage such as solid rocks, which are always devoid of free oxygen; traces of oxygen, however, can enter the upper surfaces of rocks which are in contact with air.

In exceptional cases, cavities and open fissures provide access for air to depths of several hundred meters. Man-made bore holes and mine shafts reach down two kilometers or more, but are insignificant on the scale of the biosphere.

In any case, these holes are rarely deeper than sea level; deep parts of the continents often lie below them. The bottom of Lake Baikal in Siberia, a true freshwater sea containing prolific life, is 1050 meters below sea level.[287]

It seems that life in the depths of the continents never reaches the average depth of the hydrosphere (3800 m).

It should be noted that recent research on the genesis of petroleum and hydrogen sulfide has lowered the previously-estimated limits of anaerobic life, indicating that the genesis of these phreatic minerals is biological, and that it took place at temperatures much higher than those of the Earth's surface. But even if the organisms involved were highly thermophilic (which has not been demonstrated), they could only live at temperatures of about 70° C—far from the isogeotherm of 100° C.

124 Hydrospheric life is predominant because the hydrosphere has a large volume, all of which is occupied by life, extending through a layer averaging 3800 meters in thickness, with a maximum of 10 kilometers. On land (21% of the planet's surface) the region of life has an average thickness of only a hundred meters, with a maximum of 2500. Moreover, life on land extends below sea level only in exceptional cases, while in the hydrosphere it penetrates more deeply — to 3800 meters.

* In the vast majority of cases, reports concerning free oxygen arise from errors of observation.

285 L. E. Kramarenko (1983) recently proposed the following schema of hydrobiogeochemical zonation of Earth's crust: I – the aerobic zone; II – the mixed zone; III – the anaerobic zone; IV – the zone within which bacteria are absent.

286 See Hesselman, 1917.

287 Recent measurements indicate that the surface of Lake Baikal is at an elevation of 456 meters above mean sea level, and that its maximum depth is 1,620 meters. Thus its deepest point is 1,164 meters below sea level.

Life in the Hydrosphere

125 The vital phenomena of the hydrosphere have remained unchanged in many respects since the Archean. Moreover, these phenomena have occurred only in certain regions of the hydrosphere during this entire period, despite the variability of life and changes in the oceans, and must be regarded as stable characteristics both of the biosphere and of the entire crust.

A basis for studying the mechanism of such phenomena is provided by the density of life in richly-animated oceanic regions that we shall call *living films and concentrations*. These are continuous, concentric regions, potential or actual, which can be regarded as secondary subdivisions of the hydrosphere. Maximal transformation of solar energy takes place in them. They should be included in any study of the geochemical effects of life, and of the influence of life processes, on the history of the planet.

The properties of these living films and concentrations are of interest in the following respects:

1 From the point of view of the distribution and types of living green matter, and thus, of the regions of the hydrosphere where most of the planet's free oxygen is created.

2 From the point of view of the distribution in space and time of newly created life in the hydrosphere — i.e., from the standpoint of multiplication, from which quantitative information may be gained about the laws which govern the intensity of geochemical energy and its periodic changes.

3 From the point of view of the relation of geochemical processes to the history of particular chemical elements in the Earth's crust, and thus, to the influence exerted by oceanic life upon the geochemistry of the planet. It will be seen that the functions of vital films and concentrations are diverse and specific, and that they have not changed over geologic time.

126 We have noted that the surface of the ocean is covered by a continuous layer of green life (§55). This is the field of production of free oxygen, which penetrates the whole ocean, including the deepest trenches, as a result of convection and diffusion.

Green autotrophic organisms are concentrated principally at depths less than 100 meters. Below 400 meters, for the most part only heterotrophic animals and bacteria are found.[288] While the entire surface of the ocean is the domain of phytoplankton, large primary producers such as algae and marine grasses play leading roles in certain places. They form concentrations of very

288 This contradicts Vernadsky's previous statements about what was dredged by the *Valdivia*.

different kinds, although often not differentiated as such by marine biologists. Algae and grasses develop abundantly in shallow littoral areas of the oceans, and algae also form floating masses in parts of the open sea. The Sargasso Sea is a striking example of the latter, with an area of more than 100,000 km². We will call these concentrations "littoral" and "sargassic," respectively.

Microscopic, unicellular organisms, concentrated near the ocean surface as phytoplankton, form the principal mass of green life. This is due to the great intensity of multiplication of phytoplankton, corresponding to a velocity (V) of 250 to 275 cm/sec. This value can reach as high as thousands of centimeters per second, whereas for littoral algae it is only 1.5 to 2.5 cm/sec., with a maximum perhaps ten times higher. If the occupation of the ocean by life, corresponding to the radiant energy received by the surface, depended solely on the speed V, phytoplankton should occupy an area one hundred times greater than that occupied by large algae. The distribution of the different types of oxygen producers approximately corresponds to this. Littoral algae are only found in shallow parts of the ocean.* The area of the seas† is less than 8% of the ocean surface,[289] and only a very small part of the ocean surface is covered by the larger algae and grasses. Eight percent of the surface therefore is the maximum area that the littoral plants might occupy, and in fact they cannot attain this limit. Floating concentrations of the sargassic algae play an even lesser role. The greatest mass of them, in the Sargasso Sea, covers only 0.02% of the surface area of the ocean.[290]

127 Green life is rarely visible to the naked eye in the ocean, and by no means represents the entire manifestation of life in the hydrosphere. The abundant development of heterotrophic life in the hydrosphere is rarely equaled on land. It is generally believed, probably correctly, that animals dominate the situation in the ocean and put their stamp upon all of the manifestations of life that are concentrated there.[291] This animal life could not develop, however, without the simultaneous existence of green primary producers, and its relative distribution is a result of the presence of photosynthesizing organisms. The close link between the feeding and breathing of these two forms of living matter is the factor that caused organisms to accumulate in the living films and concentrations of the oceans.

* Where great depths occur near the shore, the bed of algae occupies an insignificant area.

† That is to say, at depths less than 1000 to 1200 meters, including the deeper basins.

289 See Shokalsky, 1917.

290 There also exist floating concentrations of phyllophoran seaweeds in the Black Sea called "Zernov's Phyllophora field" (A. Lapo, written communication).

291 This interesting statement could only be true for times after the Cambrian explosion. The exact meaning of Vernadsky's term "dominate" here is unclear. If it means in terms of total biomass, it must be incorrect. If it means ecological dominance by heterotrophs, then surely it is right.

128 Living matter constitutes such a small fraction of the total mass of the ocean that sea water can be said to be mostly inanimate.[292] Even the auto- and heterotrophic bacteria, though widely dispersed, make up only an insignificant part of the ocean's weight. In this respect, they resemble the rare chemical ions of marine solutions. Large quantities of living organisms are found only in the living films and concentrations, but even here, we find only one percent of living matter by weight; the amount can reach a few percent in certain places, but only temporarily.

These living films and concentrations form regions of powerful chemical activity in which life is perpetually moving. The formations as a whole remain almost stationary, however, and are zones of stable equilibria within the changing structure of the biosphere — zones that are as constant and characteristic of the ocean as are the sea currents.

We distinguish *four stable groupings* of life in the ocean: *two films*, the planktonic and the benthic; and *two concentrations*, the littoral and the sargassic.[293]

129 The most characteristic of these living collections is the essential thin, upper layer of phytoplankton, which can be said to cover the entire ocean surface with rich green life.[294]

Sometimes the primary producers predominate in this film, but heterotrophic animals that depend upon phytoplankton for their existence are probably just as important, because of their global role in planetary chemistry.

The phytoplankton are always unicellular; the metazoans are an important part of zooplankton, and are sometimes found in larger quantities than on land. Thus, from time to time, we observe in the oceanic plankton the hard and soft roes of fish, crustacea, worms, starfish, and the like in prodigious quantities, exceeding those of all other living things. According to Hjort, the number of individuals averages between 3 and 15 microscopic phytoplankton per cubic centimeter.[295] This number rises to 100 for all the microplankton.[296] The number of unicellular phytoplankton is usually smaller than that of the heterotrophic animals, not including bacteria or nanoplankton. The planktonic film, therefore, carries hundreds, if not thousands, of microscopic individuals per cubic centimeter, each an independent center of geochemical energy (§48). This dispersed living matter cannot be less than 10^{-3} to 10^{-4} percent, by weight, of the total mass of ocean water, and is probably exceeds this amount.

292 Contemporary results support this view. Life becomes sparse in the waters near the Hawaiian Islands, where living matter exists in sea water at concentrations less than 3×10^{-6} %.

293 In the sections that follow Vernadsky is not entirely consistent in his use of terminology (A. Lapo, written communication). For instance, he calls sargassic and littoral concentrations films whereas bottom films are called concentrations. In later work (V. I. Vernadsky [W. J. Vernadsky], 1933a) he used these terms with greater precision, for example, "bottom film" becomes synonymous with "benthic film." In his *The Chemical Structure of the Earth's Biosphere and its Surroundings* (Vernadsky, 1965), Vernadsky identifies a third type of living concentration, namely, oceanic reefs.

294 This is a misapprehension on Vernadsky's part. Much of the ocean surface is a nutrient starved biological desert, unable to support abundant phytoplankton.

295 See Hjort and Gran, 1900.

296 See Allen, 1919.

This layer is tens of meters thick, and is usually located at a depth of 20 to 50 meters.[297] From time to time, the plankton rise and fall in relation to the surface. The density of individuals falls off rapidly outside this film, especially beneath it, and at depths below 400 meters individual plankters are very dispersed.

Living organisms in the mass of the ocean (average depth 3800 m, max. 10 km) thus form an extremely thin film, averaging only a few percent of the whole thickness of the hydrosphere. *From the point of view of chemistry, this part of the ocean can be considered active, and the remainder biochemically weak.*

It is evident that the planktonic film constitutes an important part of the mechanism of the biosphere. In spite of its thinness, one is reminded of the ozone screen, which is important despite the insignificant fraction of ozone concentration within it.

The area of the plankton layer is hundreds of millions of square kilometers, and its weight must be on the order of 10^{15} to 10^{16} tons.[298]

130 *The living film of the ocean bottom* is found in marine mud and in the thin layer of water adjacent to it.[299] This thin film resembles the planktonic film in size and volume, but exceeds it in weight.

It consists of two parts. The upper film, the *pelogen*,* is a region of *free oxygen*. Rich animal life exists at its surface, where metazoans play an important role, and relationships between organisms are very complex. The quantitative study of this biocenose[300] is just beginning.

This fauna is highly developed in certain places. As we have already indicated, concentrations of benthic metazoa per hectare are approximately equal to those of the vegetable metaphytes on land at their highest yield[301] (§58).

The number of benthic marine animals decreased noticeably at depths of four to six kilometers, and in the deepest trenches, macroscopic animals seem to disappear (below 7 km).

Below the benthos is the layer of bottom mud — the lower part of the bottom film. Protista are here in immense quantities, and a dominant role is played by bacteria with their tremendous geochemical energy. Only the thin, upper pelogen layer, a few centimeters thick, contains free oxygen. Below this is a thick layer of mud saturated with anaerobic bacteria and innumerable burrowing animals.

In this deeper mud, all chemical reactions occur in a highly reducing medium. The role of this relatively thin layer in the

[297] Recent work has led to a much better characterization of the "green layer" of the ocean. See Menzel and Ryther, 1960; Cullen, 1982; Trees, Bidigare and Brooks, 1986; Varela, Cruzado, Tintore and Ladona, 1992; and Venrick, McGowan and Mantayla, 1972.

[298] This figure includes the weight of the water in which the plankton live.

[299] There is a close correspondence between the biomass in the benthic film and the amount of planktonic biomass in the immediately overlying water column. In other words, marine waters with elevated planktonic biomass correspond to areas of the sea floor with large amounts of living matter in the benthic film of life. This correlation is called the "Zenkevich conformity principle" after the eminent Russian oceanographer L. A. Zenkevich (1889-1970) who discovered the effect (see Strakhov, 1978 –A. Lapo, written communication).

* This term, proposed by M. M. Solovieff, has been adopted by Russian limnologists.

[300] A biocenose is a community of organisms occupying a particular habitat.

[301] In Vernadsky's time, the bottom film had been studied only on the shelf. Vernadsky mistakenly extrapolated the shelf data to deeper regions of the ocean floor. Subsequent research has shown that in the deepest parts of the ocean the biomass of the benthos consists of less than 1 gram of living matter per square meter of the bottom. A camera placed on the sea floor in the east-equatorial region of the Pacific, timed to take photographs every four hours over a period of 202 days, captured only 35 animals on film during this period (Paul, Thorndike, Sullivan, Heezen, and Gerard, 1978; A. Lapo, personal communication).

chemistry of the biosphere is enormous (§141). The thickness of the bottom film, including the layer of mud, sometimes exceeds 100 meters, and may be much thicker. This is so, for example, in the trenches occupied by organisms like the crinoids, which seem to be very important in the chemical processes of the Earth. At the moment, we can only make a conventional estimate that the thickness of this living concentration is from 10 to 60 meters.

131 Plankton and the benthic film of life spread out over the entire hydrosphere. If the area occupied by the plankton is approximately that of the ocean itself (3.6×10^8 km), then the benthic film should be larger, since it conforms to the irregular sea floor surface.

These two films surround the hydrosphere, and are connected to the two other vital concentrations close to the oxygen-rich surface of the planet — namely, the littoral and sargassic concentrations.

The littoral vital concentrations sometimes include the entire volume of water down to the bottom film, since they are adapted to shallower regions of the hydrosphere.

The area of the littoral concentrations never exceeds ten percent of the surface area of the ocean. Their average thickness is hundreds of meters, and sometimes is as much as 500 or 1000 meters. Sometimes they form common agglomerations with the planktonic and benthic films.

Littoral concentrations in the shallower regions of ocean and seas are dependent upon the penetration of solar luminous and thermal radiation into the water, and upon the outpouring of rivers that deliver organic remains and terrestrial dust in solution and suspension. These littoral concentrations consist partly of forests of algae and marine grasses, and partly of collections of mollusks, coral, calcareous algae, and bryozoa. The overall amount of littoral living matter should be less than that in the planktonic and benthic films, since less than one-tenth of the surface area of the ocean is represented by regions having a depth less than one kilometer.

132 The *sargassic living concentrations*, which have attracted little attention until recently, appear to occupy a special place, and have been explained in a variety of ways. They differ from the planktonic film in the character of their flora and fauna, and from littoral concentrations by independence from the debris of continents and the biogenic products of rivers. In contrast to littoral concentrations, sargassic ones are found on the surface of deep regions of the ocean, having no connection with the benthos and the bottom film.

For a long time, they were considered to be secondary formations consisting of floating debris carried off from littoral concentrations by winds and ocean currents.[302] Their fixed location at definite points in the ocean was explained by the distribution of winds and currents, which under appropriate circumstances form becalmed regions. While such opinions are still popular in scientific literature, they have been refuted, at least in the case of the largest and best-studied of these formations, the Sargasso Sea.

Special flora and fauna are found there, some of which clearly have their origin in the benthos of littoral regions. It is very likely that L. Germain was correct in relating their origin to the slow adaptation of fauna and flora to new conditions;[303] i.e., to the evolution of living littoral matter during the gradual descent[304] of a continent or group of islands where the Sargasso Sea is found today.[305]

The future will show whether or not it is possible to apply this explanation to other biogenic accumulations of similar kind. In any case, it is irrefutable that this type of living concentration, with its rich population of large plants and particular types of animals, is quite different from the planktonic and benthic films. No exact measure of sargassic concentrations has been made, but the area they cover is apparently not extensive, and is certainly small in comparison with the littoral concentrations.

133 The facts show that barely 2% of the ocean is occupied by concentrations of life, and that life in the other region is highly dispersed.

These living concentrations and films exert a considerable influence on the entire ocean, particularly on its chemical composition, chemical processes, and gaseous systems. Organisms outside these vital layers do not, however, cause significant quantitative changes in the ocean. In our quantitative evaluation of life in the biosphere, we shall therefore neglect the principal mass of the ocean, and consider only four regions and their biomasses; namely, the planktonic and benthic living films and the sargassic and littoral concentrations.

134 Interruptions of multiplication occur at regularly spaced intervals in all these biocenoses. The rhythm of multiplication

302 It has been hypothesized that the living concentration of the Sargasso Sea first formed in the Miocene in the Tethyan seaway (in the vicinity of the modern Carpathian Mountains) and then subsequently migrated 8000 kilometers to the west to its present position in the Atlantic Ocean (Jerzmanska and Kotlarczyk, 1976; Menzel and Ryther, 1960).

303 See Germain, 1924, 1925.

304 This old idea, which dates back to Jules Rengade in the 1870's, was rearticulated by Edward W. Berry in 1945, in which envisaged "a laying bare of the shallow sea bottom and the direct survival and modification of some of its denizens into Land plants." This curious and incorrect idea has elements in common with Lysenkoism (1948), particularly as regards the implied inheritance of acquired characteristics.

305 This passage has a decidedly antique flavor, smacking of the legend of Atlantis, although Vernadsky is here referring to submerged land bridges or the like.

in vital films and concentrations determines the intensity of biogeochemical work on the entire planet.

As we have seen, the most characteristic feature of both oceanic living films is the preponderance of protista — organisms of small size and high speed of multiplication. Their speed of transmission of life under favorable conditions probably approaches 1000 cm/second. They also are endowed with the greatest intensity of gaseous exchange (proportional to surface area, as always), and exhibit the greatest geochemical kinetic energy per hectare (§41); in other words, they can reach the maximum density of living matter and the limit of fecundity more quickly than any other kind of organism.

The protista in the plankton are clearly different from those in the benthic living film. Bacteria predominate in the enormous mass of non-decomposed debris, derived from larger organisms, that accumulates in the benthic living film. In the planktonic film, bacteria take a second place in terms of mass relative to green protista and protozoa.

135 The protozoa are not the major form of animal life in the planktonic film. Metazoa, such as crustacea, larvae, eggs, young fish, etc., are more prominent in this region.

The rhythm of multiplication of metazoa is generally slower than that of protozoa. For these higher forms, the speed of transmission of life is fractions of a centimeter per second. For oceanic fish and planktonic crustacea, the value V does not seem to fall below a few tenths of a centimeter per second.

An enormous quantity of metazoa, often including large individuals, characterizes the benthic vital film. Metazoan multiplication is sometimes slower than that of smaller planktonic organisms. It is possible that organisms with a very low intensity of multiplication can be found in the benthic layer.

In littoral and sargassic concentrations, protista occupy second place and do not determine the intensity of geochemical processes. Metazoa and metaphytes are the characteristic forms of life in these biocenoses. Metazoa play a role that increases with depth, particularly in littoral concentrations, and they become the basic form of life in deeper regions. Their importance is apparent in the extensive colonies of corals, hydroids, crinoids, and bryozoa.

136 The progression and rhythm of multiplication are scientifically not well understood. We know only that multiplication does not continue uninterrupted, but rather in a defined and repetitive way, closely linked with astronomical phenomena. Multiplication depends on the intensity of solar radiation, on the quantity of life, and on the character of the environment.

The intensity of multiplication, which is specific to each type of organism, is proportional to the rate of migration[306] of atoms needed for the life of the organism, and to the quantity of such atoms that the organism contains. The planktonic film provides the currently best-understood example of this phenomenon.

137 As for the planktonic film, rhythmic changes brought about by multiplication correspond to annual cycles in the vital medium, and are closely connected with movements of the ocean. Tides, and changes in temperature, salinity, evaporation, and intensity of solar radiation, all have a cosmic origin.

A related phenomenon is the creative wave of organic matter in the form of new individuals which spreads across the ocean in the spring, and diminishes in the summer months. It is apparent in the annual breeding cycle of nearly all "advanced" creatures, and affects the composition of plankton. "With just as much certainty as the approach of the spring equinox and the rise in temperature, and just as precisely, the mass of plant and animal plankton in a given volume of sea water reaches its annual maximum and then again decreases."[307] Johnstone's picture is characteristic of our latitudes, but it is also true for the whole ocean, *mutatis mutandis*.

The plankton is a biocenose within which all organisms are closely connected, the most frequently observed being copepod crustacea living on diatoms, and diatoms themselves, in the North Atlantic.

One such annual rhythm is observable in the seas of northeast Europe. In February-June (for most fish, from March to April) the plankton is loaded with fish eggs in the form of hard roe. In Spring, after March, siliceous diatoms, such as *Biddulphia, Coccinodiscus*, and later some of the dinoflagellates, begin to swarm. The numbers of diatoms and dinoflagellates decrease rapidly as summer approaches, but they are soon replaced by copepods and other zooplankters. September and October bring a second, less-intensive expansion of phytoplankton in the form of diatoms and dinoflagellates. December, and particularly January and February, are characterized by an impoverishment of life and slower multiplication.

306 Vernadsky means here the horizontal migration of matter, more or less
parallel to Earth's surface.

307 See Johnstone, 1911, 1908 and 1926.

The rhythm of multiplication is characteristic, constant, and distinct for each species, and is repeated from year to year with the unchanging precision of all cosmic phenomena.

Geochemical Cycles of the Living Concentrations and Films of the Hydrosphere

138 The geochemical effects of multiplication appear in the rhythm of terrestrial chemical processes, which create specific chemical compounds in each living film and concentration. Once chemical elements have entered into the cycles of living matter they remain there forever and never again emerge, except for a small portion that become detached in the form of vadose minerals. It is precisely this fraction that creates the chemistry of the ocean. The intensity of multiplication of organisms is thus reflected in the rate of formation of vadose deposits.

The planktonic film is the principal source of the free oxygen produced by green organisms. The nitrogen compounds concentrated in it play an enormous role in the terrestrial chemistry of this element. This film is also the central source of the organic compounds created in ocean waters. Several times a year, calcium is collected there in the form of carbonates, and likewise silicon is collected in the form of opal; these compounds end up as part of the benthic film. The geological accumulations of this work can be observed in the thick deposits of sedimentary rocks,[308] in chalky limestones (nanoplankton algae, foraminifera) and in cherts (siliceous deposits of diatoms, sponges, and radiolaria).

139 The sargassic and to a certain degree the littoral concentrations are analogous to the planktonic film in terms of their respective chemical products. They, too, play a great part in the formation of free oxygen, of oxides of nitrogen and of sulfur, and of carbon-containing nitrogen and calcium compounds.[309]

It appears evident that, in these areas of concentration, magnesium enters the solid parts of organisms in significant amounts, although to a lesser degree than calcium, and immediately passes via this route into the composition of vadose minerals.[310]

Concentrations of life are much less important than plankton[311] in the history of silicon. The cyclic migration of this element in living matter is, however, very intense.

140 The history of all the chemical elements in living concentrations and films is characterized by two different processes: 1.

308 A. P. Lisitsyn (1978) has characterized a specific planktogenic subtype of carbonaceous sediment. There has long been an implicit recognition of this subtype in the West; for example, the Cretaceous Period itself (Latin *creta*, chalk) is named for the abundant chalk deposits formed at the time.

309 For example, see H. Pestana, 1985.

310 Such as dolomite.

311 Because of the use of opal (hydrated silica) in diatom, radiolarian and other planktonic protist tests.

the migration of chemical elements (specific and distinct for each element) through living matter; and 2. their escape from living matter in the form of vadose compounds. The total amount of material escaping during a short cycle of life, say one year, is imperceptible. The quantity of the elements that quit the cycle can only become perceptible after geological, rather than historical, periods of time. The masses of inert solid matter that are created in this way are, however, much larger than the weight of living matter existing at any given moment on the planet.

From this point of view, the planktonic film is very different from the littoral concentrations* of life. The vital cycle of the latter releases much greater quantities of chemical elements, and consequently leaves greater traces in the structure of the crust.[312]

These phenomena occur most intensely in the lower layers of littoral concentrations, near the bottom film, and in parts contiguous to land. The latter is characterized by the separation of *carbon* and *nitrogen* compounds and by the evaporation of *hydrogen sulfide* gas, with the consequent escape of *sulfur* from these parts of the crust. By this biochemical route, sulfates escape from lakes and saline lagoons that are formed on the borders of sea basins.

141 In littoral concentrations there is no clear boundary between the chemical reactions of the bottom mud and those at the surface, whereas in the open sea they are separated by kilometers of chemically-inert water. In shallow seas and near coasts, the boundaries disappear, and the actions of these living agglomerations merge to form regions of particularly intense biochemical work.

The intensity of such work is always high in the bottom film, where the predominant role is played by the organisms endowed with maximum geochemical energy; namely, bacteria. The chemical conditions there are strikingly different from those in other media, because of the presence of large quantities of compounds, mostly vital products, which avidly absorb the free oxygen supplied from the ocean surface. A *reducing medium* thus exists in the marine mud of the bottom film. This is the realm of anaerobic bacteria. Oxidizing reactions can only occur in the pelogen, a layer several millimeters thick in which intense oxygenating, biochemical processes produce nitrates and sulfates.

This layer separates the upper population of the benthic concentrations of life (which are analogous in their chemistry to the littoral concentrations) from that of the reducing medium of the

* The phenomena which take place in sargassic concentrations are not known precisely.

312 In particular, the limestones and pyrite-rich sediments deposited on continental margins.

bottom mud, a type of medium that is not encountered else-where in the biosphere.

The equilibrium established between the oxidizing and reducing media is constantly upset by the incessant labors of burrowing animals.[313] Biochemical and chemical reactions take place in both directions, supporting the production of unstable bodies rich in free chemical energy. It is not possible, at present, to assess the geochemical importance of this phenomenon.[314]

The principal characteristic of living bottom films is the constant deposition of rotting debris, of dead organisms descending from the planktonic, sargassic, and littoral films. This organic debris, saturated with primarily anaerobic bacteria, adds to and reinforces the reducing chemical character of the bottom films.

142 Because of the character of their living matter, the bottom concentrations of life play an absolutely essential role in the biosphere. This is important in the creation of inert matter, because the special products of their biochemical processes under anaerobic conditions are not gases, but solids, or colloids which generally become solids.

Ambient conditions are favorable for the preservation of such solids, because in this region, organisms and their remains are not subjected to the usual biochemical conditions of decomposition and putrefaction, and are rapidly shielded from the oxidizing processes that would normally cause much of their matter to end up in gaseous form. They are not oxidized or "consumed" because both aerobic and anaerobic life are extinguished at a relatively shallow depth in the marine mud. As vital remains and particles of inert, suspended matter fall from above, the lower layers of sea mud become inanimate. The chemical bodies formed by life have no time to be transformed into gaseous products or to enter into new living matter. The layer of living mud,[315] never more than a few meters in thickness, is constantly growing from above, while at its bottom life is being constantly extinguished.

The transformation of organic remains into gas, causing "disappearance," occurs only as a result of biochemical processes. In layers that do not contain life, organic debris is transformed quite differently, changing during geological time into solid and colloidal vadose minerals.

Products of the latter type of transformation are found everywhere. Surface layers of sedimentary rock several kilometers thick become metamorphised by slow chemical changes, and

313 Contemporary observations reinforce this view. According to results of recent radiocarbon investigations made by V. M. Kuptsov from the Institute of Oceanography in Moscow, a layer of mud 20 centimeters in thickness at the sediment-water interface on the sea floor could not be subjected to detailed radiocarbon analysis because of pervasive mixing of the sedimentary layers by the burrowing activities of benthic animals (A. Lapo, personal communication). On many parts of the sea floor the bottom consists of a soupy substrate consisting of a centimeter thick or thicker mantle of easily resuspended mud that results from physical reworking by burrowing animals and continuous settling out of silt and clay from ambient wave and current action (Bokuniewicz and Gordon, 1980).

314 Although it has been strongly implicated as a contributor to the trophic changes associated with the Cambrian explosion of skeletonized animals (Dalziel, 1997; McMenamin and McMenamin, 1990).

315 This "living mud" metaphor is pursued by Stolz, 1983; Stolz and Margulis, 1984; and Stolz, Margulis and Guardans, 1987.

enter the structure of massive hyperabyssal layers of phreatic or juvenile material, after encountering high temperatures in the magma envelope.[316] Later, they reenter the biosphere under the influence of the energy corresponding to these temperatures (§77, 78). They bring free energy to these deeper regions of the planet—energy previously obtained from sunlight, and changed to chemical form by life.

143 The living bottom films and contiguous littoral concentrations merit particular attention in relation to the chemical work of living matter on the planet. These chemically powerful and active regions are slow-acting, but have been uniform and equal in their effects throughout geological time. The distribution of the seas and continents on the Earth's surface give an idea of how these concentrations have shifted in time and space.

In the bottom film, the upper oxidizing part (mainly the benthos) and the lower reducing part are geochemically important, and at depths less than 400 meters they are especially so. At such depths, these layers become mixed with vital littoral concentrations, and their products are supplemented by others that are geochemically associated with photosynthetic life. (§55)

The oxidizing medium of the bottom film has clearly influenced the history of many elements in addition to oxygen, nitrogen, and carbon. This medium completely alters the terrestrial history of calcium. It is very characteristic that calcium is the predominant metal in living matter. It probably exceeds one percent by weight of the average composition of living matter, and in many organisms (mostly marine) exceeds 10% or even 20%. The action of living matter separates calcium from the sodium, magnesium, potassium, and iron of the biosphere, even though it is similar in abundance to these elements, and combines with them in common molecules in all the inert matter of the Earth's crust.[317] Calcium is separated by living processes in the form of complex carbonates and phosphates and, more rarely, oxalates, and thus remains in vadose minerals of biochemical origin with only slight modifications.

The littoral and bottom concentrations of ocean life provide the principal mechanism for formation of calcium agglomerations, which do not exist in the silicon-rich, juvenile parts of the crust, nor in its deep phreatic regions.[318]

At least 6×10^{14} grams of calcium, in the form of carbonates, is liberated each year in the ocean. There are 10^{18} to 10^{19} grams of calcium in a state of migration in the cycle of living matter;

[316] High temperatures will also be encountered without the direct association of magma because of heat of burial. Sediments with abundant fixed (organic) compounds generate oil and gas when they pass through the thermal "window" of natural gas generation (Epstein, Epstein and Harris, 1977; Harris, Harris and Epstein, 1978; Harris, 1979).

[317] This aspect of the Vernadskian research program has been carried forward by Boichenko (1986).

[318] Although calcite (plus dolomite and ankerite; see Williams, Turner and Gilbert, 1982) can form as part of unusual igneous deposits called carbonatites. Carbonatites are associated with ultrabasic alkaline rocks, particularly alkaline pyroxinites.

this constitutes an appreciable fraction of all the calcium in the crust (about 7×10^{24} grams), and a very considerable proportion of the calcium in the biosphere. Calcium is concentrated by organisms of the benthos endowed with high speeds of transmission of life, such as mollusks, crinoids, starfish, algae, coral, hydroids and others, and also by protista in the sea mud, by plankton (including nanoplankton), and lastly by bacteria,[319] which have the highest kinetic geochemical energy of all living matter.

319 See, for example, Castanier, Maurin, and Bianchi, 1984.

The calcium compounds thus released form entire mountain ranges, some of which have volumes of millions of cubic meters. The whole process, in which solar energy controls the activity of living organisms and determines the chemistry of the Earth's crust, is comparable in magnitude to the formation of organic compounds and free oxygen by decomposition of water and CO_2.

Calcium, in the form principally of carbonates, but also of phosphates, is carried by rivers to the ocean from land, where most of it has already passed through living matter in another form. (§156)

144 These oceanic concentrations of life have an influence, analogous to that described above, on the history of other common elements of the Earth's crust, certainly among these silicon, aluminum, iron, magnesium, and phosphorus. Many details concerning the distributions of elements are still obscure, but the immense importance of living films in the geochemical history of these elements cannot be doubted.

In the history of silicon, the influence of the bottom film is revealed by deposits of debris of siliceous organisms — radiolaria, diatoms, sponges — which come partly from plankton and partly from the benthos. The greatest known deposits of free silica, having volumes amounting to millions of cubic kilometers, are formed at the sediment-water interface. Free silica is inert and not susceptible to change in the biosphere. It is, however, a powerful chemical factor and a carrier of free chemical energy in the metamorphic and magmatic envelopes of the Earth because of its chemical character as a free anhydrous acid.

There can be no doubt about another biochemical action that takes place, although its importance is still hard to evaluate. This is the decomposition of alumino-silicates of kaolinitic structure by diatoms and perhaps even by bacteria, resulting on the one hand the free silica deposits mentioned above, and on the other hand in the separation of alumina hydrates. This process seems to take place not only in muddy sediment but, to judge by the experiments of J. Murray and R. Irvine, in suspensions of clay-like particles in sea water.[320] These are, themselves, the result of biochemical processes of superficial alteration of inert matter in islands and continents.

145 The largest known concentrations of iron and manganese[321] in the Earth's crust have undoubtedly resulted from biochemical reactions in the bottom layer. Such are the young Tertiary iron ores of Kertch and the Mesozoic ores of Lorraine. All evidence indicates that these limonites and iron-rich chlorites were formed in a manner closely connected with living processes.[322] Although the chemical reactions involved in this phenomenon are still unknown, the principal fact of their biochemical, bacterial character evokes no doubts. The recent works of Russian scholars like B. Perfilieff, V. Butkevich, and B. Isachenko (1926-1927) have demonstrated this.[323]

The same processes have been repeated, throughout all geological history, since Archean times.[324] In this manner, for example, the largest and oldest concentrations of iron were formed in Minnesota.[325]

Numerous manganese minerals,[326] including their greatest concentrations in the District of Kutaisi in the Transcaucasus, have an analogous character. There are transitions between iron and manganese minerals, and at the present moment analogous syntheses are occurring over considerable areas of the ocean floor. There can be no serious doubt concerning the biochemical and bacterial nature of these processes.

146 Phosphorus compounds are being deposited on the sea bottom in a similar fashion. Their connection with life processes is undoubted, but the mechanism is not known.[327]

Phosphorite deposits, which have been found in all geological periods since the Cambrian, are of biogenic origin.[328] There is no doubt that these concretionary accumulations of phosphorus, which are being formed today, for example, on a small scale near the South African coast, are related to the living bottom concentrations. Part of this phosphorus had certainly been accumulated in the form of complex phosphates concentrated in different parts of living organisms.

The phosphorus necessary to the life of organisms does not, as a rule, leave the vital cycle.[329] The conditions in which it can

320 See Murray and Irvine, 1893; and Murray, 1913.

321 Vernadsky pursued this topic further in 1934.

322 Sea-floor manganese nodules, formed of concentric layers of manganese oxides accreted around solid objects such as a shark teeth or whale ear bones (see Gupta, 1987), are also thought to be associated with bacteria (Banerjee Iyer, 1991; Janin, 1987).

323 See Perfil'ev, 1926; Butkevich, 1928; Isachenko, 1927; and Perfil'ev, 1964. See also Cowen and Bruland, 1985.

324 Again, Vernadsky expresses his uniformitarian views. Note again how Vernadsky so completely misses the oxygen crisis (the buildup of oxygen in Earth's atmosphere two billion years ago) and its relationship to banded iron formations he alludes to at the end of this paragraph. Although it would have obliterated any geochemical notion of substantive uniformitarianism, it would have boldly vindicated Vernadsky's notion of life as a geological force. How did Vernadsky, or his successors, neglect to make this important discovery? Historical accident may provide a key. Banded iron formations similar to those of the famous 2 billion year old ones of the Precambrian (such as the ones in Minnesota Vernadsky cites), complete with parallel bands of hematite, occur in the Altai Mountains of western Siberia and eastern Kazakhstan. The earliest research into the geological distribution of iron deposits of the Altai took place around 1927 and it is possible that early reports influenced Vernadsky's views of the significance of banded iron deposits before the last edition of *The Biosphere* was published in 1929. The Devonian examples of banded iron shown in Plate 59, figures 1 and 2 of Kalugin (1970) are quite similar to Precambrian examples, and might lead one to conclude that banded iron formation is a more generally geologically wide ranging type of deposit than it actually is.
In fairness to Vernadsky, even later researchers in the West have stumbled on the banded iron formation question. Cairns-Smith (1978) incor-

escape are not clear, but everything indicates that the phosphorus of organic colloidal compounds and the phosphates of body fluids are transformed into concretions and emigrate from the vital cycle along with phosphorus in calcium compounds of skeletons.

This emigration of phosphorus takes place at the death of organisms with phosphorus-rich skeletons whenever conditions prevent the usual processes of bodily alteration, and create a medium favorable for specific bacteria. However this may be, one cannot doubt the biogenic origin of these phosphoric masses, nor their close and permanent connection with the living bottom film, nor the regular occurrence of analogous phenomena throughout geological time. Great concentrations of phosphorus have been accumulated in this way, including the Tertiary phosphorites of North Africa and of the southeastern states of North America.

147 Even with our incomplete present knowledge, it is obvious that the bottom film is important in the history of magnesium and of barium, and probably of vanadium, strontium and uranium.[330] These subjects have been little-explored, as yet.

The lower part of the bottom film contains no oxygen, and is poorly studied. This region of anaerobic bacterial life contains organic compounds that have reached it following their genesis in a different, oxygen-rich medium by living organisms foreign to familiar environments. Although we can, at best, make conjectures about this obscure region. The processes occurring there must be taken into account in estimating the role of life in the Earth's crust.

Two undoubted empirical generalizations can be stated: 1. these deposits of marine mud and organic debris are important in the history of sulfur, phosphorus, iron, copper, lead, silver, nickel, vanadium, and (according to all appearances) cobalt, and perhaps other rarer metals;[331] and 2. the renewal of this phenomenon on an important scale, at different epochs in geological history, indicates its connection with specific physicogeographical and biological conditions that prevailed when marine basins slowly dried up in ancient times.

148 Certain bacteria liberate sulfur in the form of hydrogen sulfide by decomposing sulfates, polythionates, and complex organic compounds.[332] The hydrogen sulfide thus liberated enters into numerous chemical reactions and produces metallic sulfides. The biochemical liberation of hydrosulfuric acid is a

rectly postulated solution photochemistry as the source of oxygen needed to precipitate the iron as ferric deposits on the sea floor. Recent research decisively supports Preston Cloud's (1973) arguments regarding the genesis of banded iron formations (Ho Ahn and Busek, 1990). Cloud retraced the intellectual steps leading up to his banded iron formation hypothesis in a very interesting article published in (1983). See also Konhauser and Ferris, 1996; Kump, 1993; and Widdel, Schnell, Heising, Ehrenreich, Assmus and Schink, 1993.

325 See Gruner, 1924; and Gruner, 1946.

326 The manganese mineral vernadite was named in honor of V. I. Vernadsky (Chukhrov, Groshkov, Berezovskaya, and Sivtsov, 1979).

327 The process is much better understood today; see Baturin, 1989. Phosphate reaches the sea floor from the water column. Anoxic sediments return phosphate to the water column; oxic sediments retain the phosphate (see Williams, 1997, p. 90). Holland (1990) argues that because of this process, the geochemistry of phosphorus controls the levels of oxygen in the atmosphere.

328 Phosphorite deposits occur in the Precambrian as well (Kholodov and Paul', 1993; Cook and Shergold, 1984.)

329 Although it may certainly be considered "non-living" if it is in the form of dissolved orthophosphate. Cells of the alga *Anabaena flos-aquae* have been used to detect the amounts of phosphorus in aquatic ecosystems, and the sensitivity of such biotic measurements compares favorably to conventional methods of assay for dissolved orthophosphates (Stewart, Fitzgerald and Burris, 1970; Patrick, 1973).

330 See Mann and Fyfe, 1985; Neruchev, 1982; and Adriano, 1992.

331 See Adriano, 1992.

332 Such bacteria are heterotrophic, in effect respiring the oxidized sulfur compounds such as sulfate the same way we respire oxygen. (Trüper, 1982.)

phenomenon characteristic of the benthic region, and goes on everywhere in marine basins. The H_2S becomes rapidly oxidized in the upper parts of these basins, yielding sulfates which can recommence the cycle of biochemical transformations.

The biochemical genesis of compounds of some metals is not so clear. Everything shows, however, that sulfur compounds of iron, copper, vanadium, and perhaps other metals have been formed by the alteration of organisms rich in these minerals.[333] Organic matter of the marine basins probably has the property of concentrating and retaining metals from weak solutions. The metals themselves, however, may sometimes have no connection with living matter.[334]

In any case, the liberation of metals would not take place if there were no debris of life; that is, if the interstitial organic portion of the marine basin were not a product of living matter.

Such processes are observed today on a large scale in the Black Sea (genesis of iron sulfide) and are very numerous elsewhere on a lesser scale. It has been possible to establish their strong development, also, in other geological periods. Immense quantities of copper were thus liberated into the biosphere in Eurasia during the Permian and Triassic periods, originating from rich organic solutions and organisms of specific chemical composition.

149 It follows that the same distribution of life has existed in the hydrosphere throughout all geological periods, and that the manifestation of life in the chemistry of the planet has remained constant. The planktonic film and the bottom film and the concentrations of life (in any case, the littoral concentrations) have functioned throughout these periods, as parts of a biochemical apparatus that has operated for hundreds of millions of years. The continual displacements of land and sea have moved these chemically active regions from one place to another, but the study of geological deposits gives no sign of any change in the structure of the hydrosphere or its chemical manifestations.[335]

From the morphological point of view, however, the living world has become unrecognizable during this time. Since evolution has evidently not had any noticeable effect on the quantity of living matter or on its average chemical composition, morphological changes must have taken place within definite frameworks[336] that did not interfere with the manifestations of life in the chemical framework of the planet.

This morphological evolution was undoubtedly connected with complex chemical processes which must have been impor-

333 At the Oklo Site (Gabon, Africa), estuarine bacteria two billion years ago precipitated large quantities of uranium out of solution. Rising levels of free oxygen two billion years ago (the "Oxygen Crisis") caused the uranium minerals to become soluble. They were moved seaward by riverine transport. By a process of organic chelation, the ancient bacterial mats absorbed the uranium before it was diluted in the sea.

These uranium isotopes became concentrated to the point where they triggered a natural, water-mediated nuclear reaction, a reaction made possible because the proportion of easily fissionable uranium was greater two billion years ago than it is today. The toxic byproducts of this astonishing, naturally-occurring reaction are still detectable in the strata. Oklo is not a unique site; sixteen occurrences of this type are known in Precambrian strata of Africa. Based on inferences about positions of ancient continents, a search is underway for evidence of the Oklo phenomenon in South America.

Of the sixteen known Oklo and the Bangombe natural fission reactors (hydrothermally altered clastic sedimentary rocks that contain abundant uraninite and authigenic clay minerals), a number are highly enriched in organic substances (see Nagy, Gauthierlafaye, Holliger, Mossman, and Leventhal, 1993). These organic-rich reactors may serve as time-tested analogs for anthropogenic nuclear-waste containment strategies.

Organic matter apparently helped to concentrate quantities of uranium sufficient to initiate the nuclear chain reactions. Liquid bitumen was generated from organic matter by hydrothermal reactions during nuclear criticality. The bitumen soon became a solid composed of polycyclic aromatic hydrocarbons and an intimate mixture of cryptocrystalline graphite, which enclosed and immobilized uraninite and the fission-generated isotopes within the uraninite. This mechanism prevented major loss of uranium and fission products from the natural nuclear reactors for well over a billion years.

334 According to A. Lapo (written communication), this remark by Vernadsky is important. Even today,

tant on the scale of an individual, or even a species. New chemical compounds were created, and old ones disappeared with the extinction of species, without appreciable repercussions on overall geochemical effects or on the planetary manifestations of life. Even a biochemical phenomenon of such enormous importance as the creation of the skeleton of metazoa, with its high concentration of calcium, phosphorus, and sometimes magnesium, took place unnoticed in the geochemical history of these elements.[337] This is true, despite the fact that before the Paleozoic Era, these organisms probably did not have skeletons. (This hypothesis, often considered an empirical generalization, explains many important features of the paleontological history of the organic world.)

The fact that introduction of the skeleton had no effect on the geochemical history of phosphorus, calcium, and magnesium gives us reason to believe that, before the creation of metazoa with skeletons, the same compounds of these elements were produced on the same scale by protista, including bacteria.[338] The same process continues today, but its role must have been much more important and universal in the past.

If these two phenomena, despite their differences in mechanism and time of occurrence, caused the biogenic migration of the same elements in identical masses, the morphological change, important though it was, exerted no effect on the geochemical history of calcium, magnesium and phosphorus. Everything seems to show that an event of this order did actually take place in the geological history of life.

Living Matter on Land

150 The land presents a totally different picture from that of the hydrosphere.[339] It contains only one *living film*,[340] consisting of the *soil and its population of fauna and flora*. The aqueous basins are *living concentrations*[341] and must be considered separately, because they are quite distinct biochemically and biologically, and completely different in their geological effect.

Life covers the land in an almost uninterrupted film. Traces are found on glaciers and eternal snows, in deserts, and on mountain summits. In extreme cases, we can speak only of a temporary absence or scarcity of life, because in one or another form it is manifested everywhere. Spaces where life is rare constitute barely 10% of the land surface. The remainder is an integral living film.

some researchers neglect to distinguish between living and non-living organic matter in geological processes.

335 Indeed, marine salinity has remained constant for at least the last 570 million years (Hinkle, 1996).

336 This is a fascinating idea, almost completely unexplored by modern evolutionary biologists. Vernadsky implies that the morphology of organisms is constrained by, and more interestingly, must conform to, ambient geochemical constraints.

337 Not so. There is a pronouced shift across the Proterozoic-Cambrian boundary from dolomites to limestones, largely as a result of skeleton-forming organisms. Reef forming animals of the last 500 million years play an important role in carbonate geochemistry on a global scale.

338 Yet another manifestation of Vernadsky's version of substantive uniformitarianism. True, there were microbial biomineralizers in the Proterozoic, but there was nothing like the variety of different types of skeletons (in both animals and other types of organisms) after the Cambrian boundary. Vernadsky does not seem to have completely grasped the empirical generalization that there is an *evolutionary* interplay between the morphology of living matter and Earth's geochemistry, all powered by cosmic radiation.

339 Contemporary views on this subject are discussed in Dobrovolsky, 1994.

340 Vernadsky later classified subaerial and soil biocenoses as separate units, but it is not clear whether he considered them as living films or as merely living concentrations (Lapo, 1980, written communication; p. 31).

341 Vernadsky later distinguished, as separate from the aqueous basin living concentrations, the flood-plain concentrations and the coastal concentrations (Vernadsky, 1954, pp. 175-176).

151 This film is thin, extending only some tens of meters above the surface in forest areas; in steppes and fields it does not reach more than a few meters. The forests of equatorial countries with the highest trees form living films having average thicknesses from 40 to 50 meters. The highest trees rise to 100 meters or more, but are lost in the general mass of plant life and are negligible in their overall effect. Life does not sink more than a few meters into the soil and subsoil.[342] Aerobic life ranges from 1 to 5 meters, and anaerobic life to tens of meters.[343] The living film thus covers the continents with a layer that extends from several tens of meters above ground to several meters below (areas of grass). Civilized humanity has introduced changes into the structure of the film on land which have no parallel in the hydrosphere. These changes are a new phenomenon in geological history, and have chemical effects yet to be determined. One of the principal changes is the systematic destruction during human history of forests, the most powerful parts of the film.[344]

152 Obvious features of this film are the annual changes in composition and manifestation (in which we ourselves are participants) produced by the solar cycle.

The predominant organisms, in terms of quantity of matter, are the green plants, including grasses and trees. In the animal population, insects, ticks, and perhaps spiders predominate. In striking contrast to oceanic life, the heterotrophic organisms — animals — play a secondary role. The most powerful parts, the great forests of tropical countries, like the African hylea, and the northern taiga are often deserts so far as the higher animals such as mammals, birds, and other vertebrates are concerned. The animal population of these tremendous areas of green organisms consists of arthropods, which seem insignificant to us. The seasonal fluctuations in multiplication, which were observed in plankton only after extended study, are obvious in the continental film. Multiplication varies as life slows down in winter in our latitudes, and awakens in the spring. This phenomenon occurs in countless ways, from the poles to the tropics. Seasonal periods are also a characteristic of soils and their invisible life. Although this last subject has been little studied, its role in the history of the planet is, as we shall see, much greater than generally admitted.

In short, for all films in the hydrosphere and on dry land there are periods, regulated by the sun, during which fluctuations occur in the intensity of geochemical energy, and in the activity

342 Vernadsky may have been referring here only to complex forms of life such as vascular plants and animals. Vernadsky understood the depth to which bacteria reach below the soil surface (Vernadsky, 1945). This depth is now known to reach 2.8 kilometers (Anonymous, 1996b; Fredricson and Onstatt, 1996).

343 Anaerobic bacteria are now known to exist in subterranean waters at depths of several kilometers.

344 Others have recognized the influence of forest clearing on species diversity (e.g. Hutchinson, 1964), but Vernadsky was first to emphasize its geochemical consequences. With contemporary concern over clearcutting of tropical rain forest, Vernadsky's comments here are prescient.

of living matter and its "vortices" of chemical elements. *Geo-chemical processes are subject to rising and falling pulsations, although the numerical laws which govern these are not yet known.*

153 The geochemical phenomena connected with the living film on land are very different from those in the oceanic films. The emigration of chemical elements from the life cycle on land never results in concentrations of vadose minerals similar to the marine deposits, which receive millions of tons per year of calcium and magnesium carbonates (limestones and dolomites), silica (opals), hydrated iron oxides (limonites), hydrated manganese compounds (pyrolusite and psilomelane), and complex phosphates of calcium (phosphorites). (§143 et seq.) All these bodies have a marine, or at least an aqueous, origin. Chemical elements in living matter on land emigrate from the vital cycle less frequently than those in the hydrosphere (§142). After the death of the organism, the matter of which it is composed is either immediately absorbed by new organisms, or escapes to the atmosphere in the form of gaseous products. These biogenetic gases O_2, CO_2, H_2O, N_2, NH_3, are absorbed at once in the gaseous exchange of living matter.

A complete dynamic equilibrium is established, thanks to the enormous geochemical work produced by living matter on land, which after tens of millions of years of existence, leaves only insignificant traces[345] in the solid bodies of which the Earth's crust is composed. The chemical elements of life on land are in incessant motion, in the form of gases and living organisms.

154 An insignificant fraction of the solid remains (probably several million tons) escapes each year from the dynamic equilibrium of the life cycle on the land. This mass escapes in the form of finely-powdered "traces" of "biogenic organic matter", composed chiefly of compounds of carbon, oxygen, hydrogen, nitrogen, and in smaller amounts of phosphorus, sulfur, iron, silicon, etc. The whole biosphere is penetrated by this powder, of which a small, still-undetermined fraction leaves the vital cycle, sometimes for millions of years.[346]

These organic remains enter into the whole matter of the biosphere, living and inert. They are accumulated in all vadose minerals and surface waters, and are carried by rivers to the sea. Their influence on chemical reactions in the biosphere is enormous, analogous to that of organic matter dissolved in natural

345 Coal deposits might be deemed an exception to this.

346 Early successional moss ecosystems require this "powder" or bulk precipitation as their primary nitrogen source (R. D. Bowden, 1991).

waters (§93). Vital organic remains are charged with free chemical energy in the thermodynamic field of the biosphere, and because of their small dimensions easily give rise to aqueous dispersed systems and colloidal solutions.

155 These remains are concentrated in soils, and cannot be considered an absolutely inert matter.[347] Living matter in soil often reaches tens of percent by weight. Soil is the region where the maximum geochemical energy of living matter is concentrated, and is the most important biogeochemical laboratory, from the point of view of geochemical results and the development of the chemical and biochemical processes that take place in it.[348]

This region is comparable in importance to the mud layer of the living film of the ocean floor (§141), but differs from it in the importance of the oxidizing layer; this is only a few millimeters or centimeters thick in the bottom mud, but can exceed one meter in the soils. Burrowing animals are powerful factors in producing homogeneity in both regions.

The soil is a region in which surface changes of energy take place in the presence of abundant free oxygen and carbon dioxide. These gases are partly formed by living matter in the soil itself.

In contrast to the sub-aerial chemistry of the Earth, the chemical formations of the soil do not enter wholly into the living vortices of elements which, according to the picturesque expression of G. Cuvier, constitute the essence of life; they are not converted into gaseous forms of natural bodies.[349] They leave the vital cycle temporarily and reappear in another imposing planetary phenomenon, the formation of natural water and the salt water of the ocean.

The soil lives to the extent that it is damp. Its processes take place in aqueous solutions and colloids. Herein lies the chemical difference between living matter in soil and living organisms above the soil. The mechanism of water on land plays the decisive role in the former case.

156 Water on land is constantly moving in a cyclic process driven by the energy of the sun. Cosmic energy influences our planet in this way, as much as by the geochemical work of life. In the whole mechanism of the crust, the action of water is absolutely decisive, and this is most obvious in the biosphere. Not only does water constitute, on average, more than two-thirds of the weight of living matter (§109), but its presence is an

347 Vernadsky later applied the concept of bio-inert matter to complex systems such as soil, and argued that bio-inert matter is created simultaneously by living organisms and inert processes and therefore represents a dynamic equilibrium of the two. Such equilibria are found in the oceans, in almost all other waters of the biosphere, oil, soil, weathering crusts on land, etc. (See Vernadsky, 1965, p. 59. See also Perelman, 1977).

348 See Kovda, 1985.

349 See Cuvier, 1826.

absolutely necessary condition for the multiplication of living organisms and the manifestation of their geochemical energy.

Life becomes a part of the mechanism of the planet only because of water.

In the biosphere, water cannot be separated from life, and life cannot be separated from water. It is difficult to establish where the influence of water ends, and the influence of heterogeneous living matter begins. Soil quickly becomes saturated, and leaking surface waters carry away its rich organic remains in solutions or suspensions. The composition of fresh water is thus directly determined by the chemistry of the soil, and is a manifestation of its biochemistry. It follows that soil determines the essential composition of river waters where, finally, all surface waters are collected.

The rivers discharge into the sea, and *the composition of oceanic water, at least its saline part, is principally due to the chemical work of soil, and its still poorly known biocenose.*

The oxidizing character of the soil medium is important here, and accounts for the final dissolved products of the soil's living matter. Sulfates and carbonates predominate in river water, along with sodium chloride. The character of these elements in river water is directly related to their biochemistry in soil, but differs sharply from the character of the solid compounds that they form in non-living envelopes.

157 Other chemical manifestations of living matter are also related to the circulation of water on land. The life in aqueous basins is quite different from that in ground regions. Phenomena in aqueous basins are in many ways analogous to the living films and concentrations of the hydrosphere. The planktonic and benthic films and littoral concentrations are recognizable on a smaller scale. We find the processes involving both oxidizing and reducing media. Finally, the emigration of chemical elements from the vital cycle plays an important role, as does the formation of solid products which later enter into the sedimentary rocks of the crust. Here, it seems that the process of liberation of solid bodies in the biosphere is linked to the phenomenon of a reducing medium, to the rapid disappearance of oxygen, and to the disappearance both of aerobic and anaerobic protista.

In spite of these points of resemblance, the geochemical effects of life on land differ fundamentally from those in the hydrosphere.

158 This difference arises from two facts, one chemical and one physical; namely, that most of the water in the aqueous basins is fresh, and that most of these basins are shallow. Vernal pools, lakes and marshes, rather than rivers, hold most of the continental water mass, and in most cases are only deep enough to contain a single living concentration — the fresh water concentration of life. Only in the fresh water seas, of which Lake Baikal is an example, do we observe separate living films analogous to those of the hydrosphere, and these inland seas are exceptional cases.

The biochemical role of lakes is distinct from that of ocean waters primarily because the chemical compounds formed in fresh water are different. The chief product consists of carbon compounds. Silica, calcium carbonate, and hydrated iron oxides play a minor role in comparison to the deposition of carbon-bearing bodies in the bottom films of aqueous basins on land. It is here, and only here, that coal and bitumen are formed in appreciable amounts. These solid, stable, oxygen-poor bodies of carbon, hydrogen, and nitrogen are the stable forms of vadose minerals which, on leaving the biosphere, pass into other organic compounds of carbon. The carbon is set free as graphite during their final transformation in metamorphic regions.

There is no appreciable quantity of stable carbon-nitrogen substances in sea water; these are never formed in the chemistry of the ocean.[350] Whether this is an effect of the chemical character of the medium, or of the structure of the living matter concerned, we cannot say. Nor is it clear why such compounds are formed in bodies of fresh water, although the process has occurred throughout geologic times. In both cases, the phenomena are certainly connected with life processes.

These masses of organic matter provide powerful sources of potential energy — "fossilized sun-rays", according to the picturesque expression of R. Mayer. They have been enormously important in the history of man, and considerably more so in the economy of nature. An idea of their scale can be obtained from the size of the known reserves of coal.

It seems almost certain that the chief sources of liquid hydrocarbons (i.e., the petroleums) lie in fresh water concentrations.[351]

It is possible that, like many beds of coal, these basins were once close to the sea. The formation of petroleum is not a surface process; it results from the apparently biochemical decomposition of debris of organisms, in the absence of free oxygen, near the lower limits of the biosphere. The process terminates in

350 Amino acids and other nitrogenous compounds occur in sufficient quantites to permit some organisms to feed by directly absorbing them from sea water (see McMenamin, 1993).

351 Lake deposits such as the oil shales (more properly, organic marlstones) of the Green River Basin can indeed harbor considerable hydrocarbon deposits. However, the bulk of these deposits worldwide are marine, with marine deposits of the ancient Tethyan seaway (Persian Gulf) accounting for most of present global petroleum reserves.

phreatic regions. The derivation from living matter of the bulk of the petroleum is confirmed by a multitude of well-established observations.

The Relationship Between the Living Films and Concentrations of the Hydrosphere and Those of Land

159 It follows from the preceding that life presents an indivisible and indissoluble whole, in which all parts are interconnected both among themselves and with the inert medium of the biosphere. In the future, this picture will no doubt rest upon a precise and quantitative basis. At the moment, we are only able to follow certain general outlines, but the foundations of this approach seem solid.

The principal fact is that the biosphere *has existed throughout all geological periods*, from the most ancient indications of the Archean.

In its essential traits, the biosphere has always been constituted in the same way. One and the same chemical apparatus, created and kept active by living matter, has been functioning continuously in the biosphere throughout geologic times, driven by the uninterrupted current of radiant solar energy. This apparatus is composed of definite vital concentrations which occupy the same places in the terrestrial envelopes of the biosphere, while constantly being transformed. These vital films and concentrations form definite secondary subdivisions of the terrestrial envelopes. They maintain a generally concentric character, though never covering the whole planet in an uninterrupted layer. They are the planet's active chemical regions and contain the diverse, stable, dynamic equilibrium systems of the terrestrial chemical elements.

These are the regions where the radiant energy of the sun is transformed into free, terrestrial chemical energy. These regions depend, on the one hand, upon the energy they receive from the sun; and on the other upon, upon the properties of living matter, the accumulator and transformer of energy. The transformation occurs in different degrees for different elements, and the properties and the distribution of the elements themselves play an important role.

160 All the living concentrations are closely related to one another, and cannot exist independently. The link between the living films and concentrations, and their unchanging character throughout time, is an eternal characteristic of the mechanism of the Earth's crust.

As no geological period has existed independently of continental areas, so no period has existed when there was only land. Only abstract scientific fantasy could conceive our planet in the form of a spheroid washed by an ocean, in the form of the "Panthalassa" of E. Suess, or in the form of a lifeless and arid peneplain, as imagined long ago by I. Kant[352] and more recently by P. Lowell.[353]

The land and the ocean have coexisted since the most remote geological times. This coexistence is basically linked with the geochemical history of the biosphere, and is a fundamental characteristic of its mechanism. From this point of view, discussions on the marine origin of continental life seem vain and fantastic. Subaerial life must be just as ancient as marine life,[354] within the limits of geological times; its forms evolve and change, but the change always takes place on the Earth's surface and not in the ocean. It if were otherwise, a sudden revolutionary change would have had to occur in the mechanism of the biosphere, and the study of geochemical processes would have revealed this. But from Archean times until the present day, the mechanism of the planet and its biosphere has remained unchanged in its essential characteristics.[355]

Recent discoveries in paleobotany seem to be changing current opinions in the ways indicated above. The earliest plants, of basal Paleozoic age, have an unexpected complexity[356] which indicates a drawn-out history of subaerial evolution.

Life remains unalterable in its essential traits throughout all geological times, and changes only in form. All the vital films (plankton, bottom, and soil) and all the vital concentrations (littoral, sargassic, and fresh water) have always existed. Their mutual relationships, and the quantities of matter connected with them, have changed from time to time; but these modifications could not have been large, because the energy input from the sun has been constant, or nearly so, throughout geological time, and because the distribution of this energy in the vital films and concentrations can only have been determined by living matter — the fundamental part, and the only variable part, of the thermodynamic field of the biosphere.

But living matter is not an accidental creation. Solar energy is reflected in it, as in all its terrestrial concentrations.

We could push this analysis further, and examine in greater depth the complex mechanism of the living films and concentrations, and the mutual chemical relationships which link them together. We hope to return at a later time to problems of homo-

352 See Kant, 1981.

353 See Lowell, 1909.

354 A fascinating insight, and apparently correct, assuming that Vernadsky refers here primarily to bacteria (Schwartzman and Volk, 1989). A. I. Perelman felt that Vernadsky was referring to more complex organisms; if so, Vernadsky was led astray by his uniformitarian leanings.

355 See Cloud, 1973, p. 1135.

356 This is indeed true, especially for the Silurian plant *Baragwanathia*. The complexity, however, is better explained by the onset of symbiosis of plants with fungi than by a long stretch of prior evolution (McMenamin and McMenamin, 1994).

geneous living matter and to the structure of living nature in the biosphere.[357]

●

357 Vernadsky did indeed return to these problems in book manuscripts which were not published until after his death, and then only in Russian (Vernadsky, 1965). A good number of Vernadsky's later works are treatments of his biosphere concept: Vernadsky, 1940; 1980; 1991; 1992; and 1994.

Appendix I

Vladimir Ivanovich Vernadsky (1863-1945)
A Biographical Chronology
Compiled by Jacques Grinevald

1863
March 12 (February 28, old style) Born in St Petersburg, Tsarist Russia.

1868
The family moves to Kharkov, Ukraine.

1873
Gymnasium. Much influenced by his uncle E. M. Korolenko (1810-80), an encyclopedist autodidact and nature-lover.

1875
Publication of *Die Enstchung der Alpen* by Eduard Suess (first mention of the "biosphere.")

1876
Back to St. Petersburg. His father, Ivan Vasslievich Vernadsky (1821-1884), a professor of political economy (Kiev, Moscow) and politically active in the liberal movement, manages a bookshop and a printing house. Vladimir will be a great reader in many languages.

1881
Faculty of Physics and Mathematics (Section of Natural Sciences), St. Petersburg University. Student of the great chemist Dmitri Mendeleyev (1834-1907), and Vasili V. Dokuchaev (1846-1903), the founder of pedology, soil science. Dokuchaev, indebted to Humboldtian science, has been the father of a large naturalist school, including S. N. Winogradsky [Vinogradsky] (1856-1946), V. Agafonoff (1863-1955), G. F. Morozov (1867-1920), K. D. Glinka (1867-1927), B. B. Polynov (1867-1952), and L. S. Berg (1876-1950), and especially V. Vernadsky (who also created a large scientific school).

1883
Elected a member of the Mineraological Society (St. Petersberg). Publication of *Das Antlitz der Erde* by Eduard Suess.

1886
Married Natalya E. Staritskaya (1860-1943). One year later, birth of their son George Vernadsky (emigrated in 1921; professor at Yale 1927, died in 1973, USA).

1888

Vernadsky obtains a scholarship for two years of advanced studies in Western Europe. Crystallography and mineralogy at Munich with Paul Groth (1843-1927). Friendship with Hans Driesch (1867-1941), Haeckel's graduate student and later famous as a controversial vitalist philosopher of organicism. Geological trip in the Alps with Karl von Zittel (1839-1904). Attended IVth International Geological Congress in London. Elected corresponding member of the British Association for the Advancement of Science. During an expedition in Wales, he meets Alexi P. Pavlov (1845-1929), who invites him to teach at Moscow University.

1889

First stay in Paris. Mineralogy at the laboratory of Ferdinand Fouqué (1828-1904), at the Collêge de France, together with Agafonoff and Alfred Lacroix (1863-1948), later Secrétaire perpetuel of the Academie des Sciences (since 1914). Thermodynamics and physical chemistry with Henry Le Chatelier (1850-1936), at the Ecole des Mines. Crystallography at the Sorbonne with Pierre Curie (1859-1906), discovering the problem of symmetry and dissymetry. Dokuchav's representative at the Exposition Internationale of Paris. Elected member of the Société française de minéralogie.

1890

Begins Master's thesis, Moscow University. Returns to Paris.

1891

Master's dissertation: "On the sillimanite group and on the role of the alumina in the silicates." Begins his twenty year professorship in mineralogy and crystallography at Moscow University.

1896

Sent on research mission to Europe (Germany, Switzerland, France). Henri Becquerel discovers radioactivity.

1897

Doctoral thesis, Moscow University.

1898

Extraordinary Professor. Birth of their daughter Nina (eventually emigrated to USA).

1902

Ordinary Professor. Lectures on the development of "a scientific world view" emphasizing the need for an unified view of nature.

1903

The Fundamentals of Crystallography. Begins his association with his favorite student Aleksandr E. Fersman (1883-1945), later a leading Soviet geochemist. The Nobel Prize in physics is shared by Henri Becquerel (1852-1908) and the Curies for the discovery of radioactivity.

1905
First democratic revolution in Russia. Founding-member of the liberal Constitutional-Democratic Party (KD). Member of its Central Committee (from 1908 to 1918.)

1908
Elected extraordinary member of the Academy of Sciences. First part of his multi-volume *Descriptive Mineralogy*. At a British Association meeting, in Dublin, he is attracted to geological implications of radioactivity by John Joly (1857-1933). Publication of *Die Energie*, by Wilhem Ostwald (1853-1932), Leipzig. Ostwald's energetism, adopted by Mach's Russian disciples, including A. Bogdanov, is attacked by Lenin's *Materialism and Empirio-Criticism*, future gospel of Stalinist epistemology.

1909
Reads *The Data of Geochemistry* by Frank W. Clarke (1847-1934). He decides to turn to geochemistry.

1910
Visits Marie Curie Sklodowska (1867-1934) in Paris, and proposes to organize an "international radiography of the earth's crust."

1911
A large group of Moscow University professors, including Vernadsky, resigns in protest against the repressive policy of the tsarist Minister of Education. Returns to St. Petersburg. Visits the great geologist Eduard Suess (1831-1914), President of the Imperial Academy of Sciences, Vienna.

1912
Full member of the Academy of Sciences, St. Petersburg.

1913
XIIIth International Geological Congress in Canada; travelling also in USA, visiting several laboratories, including the Geophysical Laboratory of the Carnegie Institution of Washington.

1914
World War I. Russia is attacked by Germany. First use of the term "biosphere" in Vernadsky's published work.

1915
Founder and Chairman (until 1930) of the Commission for the Study of Natural Productive Forces (KEPS), directed to organize "scientific, technical, and social forces for more effective participation in the war effort." Publication of *Die Entsehung der Kontinente und Ozeane* by Alfred Wegener.

1916

Chairman of the Scientific Board of the Ministry of Agriculture. Einstein proposes general theory of relativity.

1917

The February Revolution. Collapse of the Tsarist regime. Member of Kerensky's Provisional Government, as Assistant to the Minister of Education. In Summer, afflicted by tuberculosis, he moves to Ukraine, where he possessed a family *dacha*. He begins writing a long manuscript on *Living Matter* (not published until 1978). The October Revolution. Russian civil war.

1918

He resigns from his party, feeling himself "morally incapable of participating in the civil war." Founding member—together with several promising scientists, including Ivan I. Schmalhausen (1884-1963)—and first President of the Ukrainian Academy of Sciences, Kiev. Lives and works in secret outside Kiev, at the Biological Research Station near Starosele on the Dnieper. Theodosius Dobzhansky (1900-75), later the famous evolutionary biologist, who emigrated (in December 1927) to the United States, is one of his research assistants (1918-19).

1920

The Vernadskys move to Crimea. Like many other anti-Bolshevik scientists, Vernadsky takes refuge as professor at the Tauride University, Simferopol, under the protection of General Wrangel's Army. He is elected Rector. This position is bright but short lived. The Vernadskys are also helped by Hoover's American Relief Administration (ARA).

1921

The White Armies are unable to resist to the Red Army. The evacuation commanded by General Wrangel, includes the Vernadskys. But only George, Venadsky's son, accepts evacuation (first emigrated to Prague). Vernadsky, his wife, and daughter are arrested by the Cheka, and brought back to Moscow. Thanks to Lenin himself, they are soon liberated. Founding father and Chairman of the Commission on the History of Knowledge, Academy of Sciences.

1922

Petrograd. The Radium Institute is founded under the direction of Vernadsky (until 1939), Fersman as deputy chairman and Vitali G. Khlopin (1890-1950) as secretary (director in 1939). At the invitation from the Rector of the Sorbonne, Paul Appell (1855-1930), and with an official scholarship (for one year) from his Academy, Vernadsky and his wife move to France, via Prague. As "Professeur agréé de l'Université de Paris," Vernadsky is invited to give lectures on "Geochemistry" (Winter 1922-23). Works at the Muséum d'histoire naturelle (A. Lacroix), and at the Institut du Radium (Marie Curie). In May, Vernadsky is received by Henri Bergson (1859-1941), then President of the Commission internationale de la coopération intellectuelle of the League of Nations. December 30, 1922: creation of the USSR (collapse in December 1991).

1923

Meeting of the British Association at Liverpool, where he is impressed by Paul Langevin (1872-1946) and Niels Bohr (1885-1962). His "plea for the establishment of a bio-geochemical laboratory" is published in Liverpool. His academic position in France is again extended for one year.

1924

La Géochimie, Félix Alcan: Paris. Receives a financial support from the Rosenthal Foundation for measuring biogeochemical energy. Many discussions with Pierre Teilhard de Chardin (1881-1955) and Edouard Le Roy (1870-1954). The trio invent the concept of "the noösphere." Publication of *The Origin of Life* by Alexander I. Oparin (1894-1980). His Academy urges him to return in Russia. Death of Lenin.

1925

"L'autotrophie de l'humanité," *Revue générale des Sciences* (September 15-30); "Sur la portée biologiques de quelques manifestations géochimiques de la vie," *Revue générale des Sciences* (May 30). The celebration of the 200th anniversary of the Soviet Academy of Sciences: the name of Vernadsky is omitted—probably a political warning. Lack of permanent funding from the West for his biogeochemical lab project, moral obligation to his friends, deep patriotism, optimism about the Soviet science policy, and loyalty to his beloved Academy forces him to return to his native country, now the USSR under the Soviet regime. Departs from Paris in December. He stays first in Prague, where his book *Biosfera*—mainly written in France—was finished. Stalin consolidates his power.

1926

Returns to Leningrad in March, with his wife, leaving George (teaching at Charles University in Prague) and Nina (now Dr. in medicine) abroad. Publication—2,000 copy first printing—of *Biosfera*. He organizes within the Academy of Sciences the Department of Living Matter. Again heads the KEPS, until its reorganization in 1930, and the Radium Institute. Elected to the Czech and Serbian Academies of Sciences, Société géologique de France, German Chemical Society, German and American Mineralogical Societies. Publication of *Holism and Evolution* by Jan Christiaan Smuts (1870-1950), the famous South African General.

1927

Thoughts on the Contemporary Significance of the History of Knowledge. Three-month tour in Western Europe. The Soviet Science Week in Berlin. Helps to create the Dokuchaev Soil Institute, directed by Glinka, then Polynov. His son George is appointed professor of Russian history at Yale University.

1928

"Le bactériophage et la vitesse de transmission de la vie dans la biosphère," *Revue générale des Sciences* (Mars 15). Elected corresponding member of the Académie des Sciences, Paris (Section of Mineralogy). His department of Living Matter (within KEPS) is reorganized into the Biogeochemical Laboratory (BIOGEL)—after World War II, the Vernadsky Institute of Geochemistry and Analytical Chemistry, Moscow, with Aleksandr P. Vinogradov (1895-1975) as first director.

1929

La Biosphère, Paris, Félix Alcan. Serious ideological assault begins on the Academy of Sciences: Vernadsky is the leader of an unsuccessful resistance against the Communist Party's progressive takeover of the Academy.

1930

Geochemie in ausgewahlten Kapiteln, translated from the Russian by Dr. E. Kordes, Leipzig, Akademische Verlagsgesellschaft (*Die Biosphäre*, Leipzig, 1930, quoted many times in Kordes's edition, was apparently never published). "L'étude de la vie et la nouvelle physique," *Revue générale des Sciences* (December 31). As are many conservationists and ecologists, Vladimir V. Stanchinsky (1882-1942), who is indebted to Vernadsky for his energetic and holistic approach of natural systems, is attacked by I. I. Prezent (1902-67), the Bolshevizer of biology and ally of T. D. Lysenko (1898-1976). The Commission on the History of Sciences, is transfomed into the Institute of the History of Science and Technology, Vernadsky is replaced as director by Nikolai Bukharin (1888-1938).

1931

Second International Congress of the History of Science and Technology, London; marked by the Marxist contributions of the official Soviet delegation, led by N. Bukharin (including A. Ioffe, N. Vavilov, B. Hessen, and, of course, not Vernadsky.)

1932

"Sur les conditions de l'apparition de la vie sur la terre," *Revue générale des Sciences*. Visites his Norwegian colleague Victor Moritz Goldschmidt (1888-1947), considered the founder of modern geochemistry, in Göttingen, Germany. Travels to Paris. His Radium Institute decides to build a cyclotron, which begins operation in the late 1930s with Igor Kurtchatov (1903-60), later the chief scientist of the Soviet atomic bomb program (secretly initiated in 1942, without Vernadsky).

1933

Invited to the University of Paris: two conferences (December 19 and 22) on radiogeology at Marie Curie's Radium Institute. Japanese translation of Vernadsky's *Geochemistry*.

1934
History of Natural Waters (in Russian). His friend and collaborator
Boris L. Lickov (1888-1966) is arrested and deported. Their corre-
spondance continues (published in 1979-80, but still in censored
form). "Le problême du temps dans la science contemporaine,"
Revue générale des Sciences; also published in booklet form. Head
of the Commission on Heavy Water (transformed into a Commission
on Isotopes in 1939) created by the Academy of Sciences.

1935
Vernadsky moves to Moscow, because of the transfer of the Academy
of Sciences of the USSR. Death of Karpinsky, Vernadsky's friend and
"bourgeois" president of the Academy (elected in 1917). Last travel in
France and abroad. *Les Problêmes de la radiogéologie*, Paris.
Problems of biogeochemistry. Increasing difficulties with publication
of his non-technical works.

1937
On the boundaries of the biosphere, Moscow, Academy of Sciences,
"Geological Series." International Geological Congress, Moscow.
Proposes an international commission for measuring geological time
by radioactive methods. Moscow show trials begin.

1938
Goethe as a Naturalist (not published until 1946). *Scientific Thought
as a Planetary Phenomenon* (not published until 1977). The Institute
of the History of Science and Technology is closed after Bukharin's
execution.

1939
World War II. His longtime friend D. Shakhovskoi is arrested (and dies
in prison the following year).

1940
Biogeochemical Essays, 1922-1932, Academy of Sciences of the USSR
(in Russian). He begins writing his major work *The Chemical Structure
of the Earth's Biosphere and its Surroundings*, never completed, not
published until 1965 (only in Russian). After receiving news from his
son George (a *New York Times* clipping of May 5 on nuclear research),
Vernadsky writes a letter (July 1) about the national need—"despite
the world military situation"—for an urgent program in atomic energy to the geophysicist Otto Yu. Schmidt (1891-1956), vice-president
of the Academy of Sciences and close to Stalin. Vernadsky, together
with his close associates Khlopin and Fersman are not unaware of
military implications of the technical use of energy within the atom,
but their main concern was about long-term energy needs of
humankind. Vernadsky urges the Soviet Academy to create a
Commission on "the uranium problem;" established in July, with the
physicist Abram F. Ioffe (1880-1960) as chairman, and Khlopin vice-
chairman.

1941

The Nazi German invasion of the USSR. Evacuated, along with other elderly academicians, to the climatic station of Borovoe, Kazakhstan. While his Kiev friend Schmalhausen is writing *Factors of Evolution* (published in 1946), Vernadsky continues to write *Chemical Structure of the Earth's Biosphere and its Surroundings*. Victim of the rising Lysenko's dogma, N. Vavilov is arrested and dismissed from all his posts (sent to a concentration camp, he dies in prison in 1943).

1943

Returns to Moscow. For his 80th jubilee, Vernadsky is officially honored with State Prize of USSR. He writes "Some words on the noösphere," published in Russia in 1944 (in USA in January 1945). After the death of his wife (February), Vernadsky returns to Moscow. He expresses the opinion that after the war the Soviet scientists need to enter into much closer contact with the American scientists.

1944

Problems of Biogeochemistry, II, translated by George Vernadsky, edited and condensed by G. E. Hutchinson, published in the *Transactions of the Connecticut Academy of Arts and Sciences*. Publication of *What Is Life?* by Erwin Schrödinger.

1945

(January 6) Death after a brain hemorrhage. Buried at the Novodevichye cemetary.

Appendix II

Vernadsky's Publications in English
Compiled by A. V. Lapo

"A plea for the establishment of a bio-geochemical laboratory." *The Marine Biological Station at Port Erin (Isle of Man) Annual Report, Transactions of the Liverpool Biological Society*, 1923, v. 37, pp. 38-43.

"Isotopes and living matter." *Chemical News*, 1931, v. 142 (3692), pp. 35-36.

"Biogeochemical role of the aluminum and silicon in soils." *Comptes Rendus (Doklady) de l'Academie des Sciences de l'USSR*, 1938, v. 21 (3), pp. 126-128.

"On some fundamental problems of biogeochemistry." *Travaux du Laboratoire Biogeochemique de l'Academie des Sciences de l'USSR*, 1939, v. 5, pp. 5-17.

"Problems of Biogeochemistry, II. The fundamental matter-energy difference between the living and the inert natural bodies of the biosphere." *Transactions of the Connecticut Academy of Arts and Sciences*, 1944, v. 35, pp. 483-517.

"The biosphere and the noösphere." *American Scientist*, 1945, v. 33 (1), pp. 1-12; republished: *Main Currents In Modern Thought*, 1946, April, pp. 49-53.

The Biosphere. (Abridged edition) Synergistic Press, Inc., 1986, 82 pgs.

The Biosphere. (Complete annotated edition) Nevraumont/Copernicus/Springer-Verlag, 1997, 192 pgs.

Bibliography

Adams, M. B., ed. 1994. *The Evolution of Theodosius Dobzhansky: Essays on His Life and Thought in Russia and America*. Princeton: Princeton University Press.

Adriano, D. C. 1992. *Biogeochemistry of Trace Metals*. Boca Raton, Florida: Lewis Publishers.

Allen, E. R. 1919. "Some conditions affecting the growth and activities of *Azotobacter chroococcum*." *J. Marine Biolog. Assoc., Plymouth, N. S.* v. 12. p. 7.

Anonymous. 1996a. "Witches' brew of weird bugs." *Frontiers, Newsletter of the National Science Foundation* October, p. 2.

———. 1996b. "Rock-eating slime." *Discover* v. 17, pp. 20-21.

Arrhenius, S. 1896. "On the influence of carbonic acid in the air upon the temperature of the ground." *Philosophical Magazine, 5th series* v. 41, pp. 237-276.

———. 1915. *Quantitative Laws in Biological Chemistry*. London: G. Bell and Sons.

Asmous, V. C. 1945. "Obituary—Academician V. I. Vernadsky 1863-1945." *Science* v. 102, pp. 439-441.

Austin, J. H. 1978. *Chase, Chance and Creativity*. New York: Columbia University Press.

Awramik, S. M., J. W. Schopf, and M. R. Walter. 1983. "Filamentous fossil bacteria from the Archean of Western Australia." *Precambrian Research* v. 20, pp. 357-374.

Backlund, H. G. 1945. "V. I. Vernadski 12/3 1963-6/1 1945 och A. E. Fersman 8/11 1883-20/5 1945; en studie i mineralogiens renaessans." *Geologiska Föreningens i Stockholm Förhandlingar* v. 67, pp. 534-548.

Baer, K. E. von. 1828. *Entwicklungsgeschichte der Thiere: Beobachtung und Reflexion*. Königsberg: Bornträger.

———. 1864-1876. *Reden gehalten in wissenschaftlichen Versammlungen und kleinere Aufsätze vermischten Inhalts*, zwei Bände. St. Petersburg: H. Schmitzdorff.

———. 1876. *Studien aus der Geschichte der Naturwissenschaften*. St. Petersburg: H. Schmitzdorf.

Bailes, K. E. 1990. *Science and Russian Culture in an Age of Revolutions: V. I. Vernadsky and His Scientific School, 1863-1945*. Bloomington and Indianapolis: Indiana University Press.

Banerjee, R., and S. D. Iyer. 1991. "Biogenic influence on the growth of ferromanganese micronodules in the Central Indian Basin." *Marine Geology* v. 97, pp. 413-421.

Bargagli Petrucci, G. 1914. "Studi sulla flora microscopia della regione boracifera toscana. V. L'ossidazione biologica dell'idrogeno solforato." *Nuovo Giorn. Bot. Ital. Firenze* v. 21, n. S, pp. 267-278.

Barlow, C., and T. Volk. 1990. "Open systems living in a closed biosphere: a new paradox for the Gaia debate." *BioSystems* v. 23, pp. 371-384.

Barrow, J. D., and F. J. Tippler. 1986. *The Anthropoloical Cosmological Principle*. New York: Oxford University Press.

Bastin, E. S. 1926. "The presence of sulphate reducing bacteria in oil field waters." *Science* v. 63, pp. 21-24.

Baturin, G. N. 1989. "The Origin of Marine Phosphorites." *International Geol. Review* v. 31, pp. 327-342.

Becquerel, P. 1910. "Recherches expérimentales sur la vie latentedes spores des mucorinées." *Comptes Rendu de Académie des Sciences, Paris* v. 150, pp. 1437-1439.

Bergson, H. 1975. *Creative Evolution* [Reprint of 1944 edition. Translated by A. Mitchell]. Westport, CT: Greenwood Publishing Group.

Bernard, C. 1866. *Leçons sur les proprietes des tissues vivants*. Paris: Bailliere.

————. 1878a. *La science experimentale*. Paris: J. B. Baillere and Son.

————. 1878b. *Leçons sur les phénomènes de la vie communs animaux at aux végétaux*. Paris: Librairie J. -B. Baillière [1966, new edition with a forward by G. Canguilhem. Paris: Vrin].

Berry, E. W. 1945. "The beginnings and history of land plants." *Johns Hopkins University Studies in Geology* v. 14, pp. 9-91.

Besredka, A. M. 1925. *Mestny Immunitet [Local Immunity]*. Paris, Soc. d'editions Sci. France-Russie [Later edition is Besredka, A. M. 1937. *Les immunités locales*. Paris: Masson et Cie.].

Bogachev, V. V. 1927. "Iskopaemaya voda i bakterial'nye protsessy v produktivnoy tolshche Apsheronskogo poluostrova" ["Ancient aquifer water and bacterial processes in productive strata of the Apsheron peninsula"]. *Azerbidzhan. Neftyanoe Kozyaistvo [Azerbaijan Oil Economics]* n. 2, pp. 48-53.

Bogdanov, A. A. 1922 [A. A. Malinovsky]. *Tektology: The Universal Organizational Science* [3rd ed.]. Moscow.

————. 1980 [A. A. Malinovsky]. *Essays in Tektology: The General Science of Organization* [Tranlated by G. Gorelik. 2nd ed., 1984]. Seaside, CA: Intersystems Publications.

Bohr, N. 1934. *Atomic Theory and the Description of Nature*. Cambridge: Cambridge University Press.

Boichenko, E. A. 1986. "Uchastiie metallov v evolutsii biogeokhimicheskikh protsessov biosfery" ["Participation of metals in the evolution of biogeochemical processes of the biosphere"]. *Geokhimiya* n. 6, pp. 869-873.

Bokuniewicz, H. J., and P. B. Gordon. 1980. "Sediment transport and deposition in Long Island Sound." *Advances in Geophysics* v. 22, pp. 69-106.

Borkin, L. Ya. 1983. "Problema mnogo- i polifilii v evolyutsionnoi teorii" ["The problem of monophyly and polyphyly in evolutionary theory"]. In *Rasvitie evolyutsionnoi teorii v SSSR [The development of evolutionary theory in the USSR]*, pp. 405-421. Leningrad: Nauka.

Bottomley, W. B. 1917. "Some effects of organic growth-promoting substances (auximones) on the growth of *Lemna minor* in mineral culture solutions." *Proc. Roy. Soc. Lond.* v. 89, n. B, pp. 481-507, June 1.

Boussingault, J. B. 1860-84. *Agronomie, chimie agricole et physiologie* [2nd ed.]. Paris: Mallet-Bachelier.

Boussingault, J. B., and J. B. Dumas. 1841. "Recherches sur la veritable constitution de l'air atmosphérique." *Comptes Rendus* XII, pp. 1005-1025.

————. 1844a. *Chemical and Physiological Balance of Organic Nature*. New York: Saxton and Miles.

————. 1844b. *Essai de statique chimique des êtres organisés*. Paris: Fortin et Masson.

Bowden, R. D. 1991. "Inputs, outputs and accumulations of nitrogen in an early successional moss (*Polytrichum*) ecosystem." *Ecological Monographs* v. 61, pp. 207-223.

Bowers, R. W., ed. 1965. *Meteoritica, Volume 23*. New York: Taurus Press [A collection of articles edited by V. G. Fesenkov and Ye. L. Krinov, translated by Spectrum Translation and Research, Inc.].

Bowler, P. J. 1992. *History of Environmental Sciences*. New York: W. W. Norton.

Bridgman, P. W. 1925. "The compressibility of several artificial and natural glasses." *American Journal of Science, series 5* v. 10, pp. 359-367.

Broeker, W. S. 1996. "The Biosphere and Me." *Today* GSA [Geologicial Society of America] v. 6, n. 7, pp. 1-7.

Brown, H. T., and F. Escombe. 1898. "Note on the influence of very low temperatures on the germinative power of seeds." *Proceedings of the Royal Society* v. 62, pp. 160-165.

———. 1900. "Static diffusion of gases and liquids in relation to the assimilation of carbon and translocation in plants." *Philosophical Transactions of the Royal Society of London* v. 193, pp. 223-292.

Brown, R. F. 1975. *Organic Chemistry*. Belmont, California: Wadsworth Publishing Company.

Budyko, M. 1986. *The Evolution of the Biosphere* [Translated from Russian]. Dordrecht: Reidel.

Buffon, G. L. L. 1792. *Buffon's Natural history: containing a theory of the earth, a general history of man, of the brute creation, and of vegetables, minerals, &c.* Volumes 1-10. London: J. S. Barr.

Bünning, E. 1989. *Ahead of his Time: Wilhelm Pfeffer. Early Advance in Plant Biology* [Translated from German by H. W. Pfeffer, 1975]. Ottawa: Cartelon University Press.

Bunyard, P., ed. 1996. *Gaia in Action*. Edinburgh: Floris Books.

Butkevich, V. S. 1928. "Obrazovanie morskikh zhelezomargantsveykh otlozheny i uchastvuyushchie v nem mikroorganizmy" ["Formation of marine iron-manganese coatings and the microbes that formed them"]. *Trudy Morskogo nauch.-issled. inst.* [*Proceedings of the Sea Research Institute*] v. 3, pp. 1-150.

Cairns-Smith, A. G. 1978. "Precambrian solution photochemistry, inverse segregation, and banded iron formations." *Nature* v. 276, pp. 807-809.

———. 1991. *Seven Clues to the Origin of Life*. Cambridge: Cambridge University Press.

Cannon, W. B. 1932. *The Wisdom of the Body*. New York: W. W. Norton & Co.

Canter, L. W. 1996. *Nitrates in Groundwater*. Boca Raton, Florida: Lewis Publishers.

Case, T. J. 1997. "Natural selection out on a limb." *Nature* v. 387, pp. 15-16.

Castanier, S., A. Maurin, and A. Bianchi. 1984. "Participation bactérienne en la précipitation du carbonate." *Comptes Rendu de Académie des Sciences, Paris* t. 299, ser. II, pp. 1333-1336.

Chamberlain, T. C. 1965. "The method of multiple working hypotheses." *Science* v. 148, pp. 754-759 [First published is 1890. "The method of multiple working hypotheses." *Science* v. 15, pp. 92-96.].

Chapelle, F. H., and P. M. Bradley. 1996. "Microbial acetogenesis as a source of organic acids in ancient Atlantic coastal plain sediments." *Geology* v. 24, p. 925-928.

Chardin, P. T. 1955-1966. *Oeuvres* [13 vols.] Paris: Editions du Seuil.

———. 1957a. "La Face de la Terre." [*Etudes* v. 5, n. 20, décembre 1921]. Reprinted in *La Vision du Passé, Oeuvres* t.3, pp. 411-67. Paris: Editions du Seuil [The Vision of the Past. 1966. Translated by J. M. Cohen.] London: Collins.

———. 1957b. "L'Histoire Natuelle du monde. Réflexions sur la valeur l'avenir de la Systématique." [*Scienta* (Revue Internationale de Synthèse Scientifique) v. 37, n. 153, janvier 1925, pp. 15-24.] Reprinted in *La Vision du Passé, Oeuvres* t. 3, pp. 143-157. Paris: Esitions du Seuil [*The Vision of the Past*. 1966. Translated by J. M. Cohen. London: Collins].

———. 1971. *L' Oeuvre scientifique* [11 vols]. Textes réunis et édités par Nicole and Karl Schmitz-Moormann, Préface de Jean Piveteau. Olten and Freiburg im Brisgau: Wlater-Verlag.

Chevreul, M. E. 1824. *Considerations générales sur l'analyse organique et sur ses applications*. Paris and Strasbourg: F. F. Levrault.

Cholodny, N. G. 1926. *Die Eisenbakterien. Beitrage zu einer Monographie*. Jena, Germany: Verlag G. Fischer.

Christensen, H. R. 1915. "Studen Uber Den Einflub der Bodenbeschaffenheit auf das bakterienleben und den stoffumsatz im erdboden." *Centralblatt für Bakt.* Abt. II, n. 1/7, bd. 43, pp. 1-166.

Christensen, H. R., and O. H. Larson. 1911. "Untersuchungen uber methoden zur bestimmung des kalbedurfisses des bodens." *Centralblatt für Bakt* Abt. II, bd. 29, pp. 347-380.

Chukhrov, F. V., A. I. Groshkov, V. V. Berezovskaya, and A. V. Sivtsov. 1979. "Contributions of the mineralogy of authigenic manganese phases from marine manganese deposits" *Mineralium deposita* v. 14, pp. 249-261.

Clark, P. 1976. "Atomism versus thermodynamics." In *Method and appraisal in the physcal sciences: The critical background to modern science, 1800-1905*, Howson, C., ed., pp. 41-105. Cambridge: Cambridge University Press.

Clarke, F. W. 1908. *The Data of Geochemistry*. Washington, D.C.: Bulletin of the U.S. Geological Survey.

Cloud, P. 1973. "Paleoecological significance of the banded iron formation." *Economic Geology* v. 68, pp. 1135-1143.

———. 1983a. "The Biosphere." *Scientific American* v. 249, pp. 132-144.

———. 1983b. "Banded Iron-Formation—a gradualist's dilemma." In *Iron-Formation: Facts and Problems*, Trendall, A. F., and R. C. Morris, eds., pp. 401-416. Amsterdam: Elsevier.

———. 1988. "Gaia modified." *Science* v. 240, p. 1716.

Colic, K., S. Petrovic, H. Sunkel, M. Burda, M. Bielik, and V. Vyskocil. 1994. "Similarities of and differences between three mountain belts at the border of the Pannonian Basin due to the behaviour of the Mohorovičič discontinuity." *Studia Geophysica et Geodaetica* v. 38, pp. 131-139.

Compton, P. 1932. *The Genius of Louis Pasteur*. New York: Macmillian Co.; Pasteur 1876.

Cook, P. J., and J. H. Shergold. 1984. "Phorphorus, phosphorites, and skeletal evoltion at the Precambrian-Cambrian boundary." *Nature* v. 308, pp. 231-236.

Cowan, G. A.. 1976. "A natural fission reactor." *Scientific American* v. 235, pp. 36-47.

Cowen, J. P., and K. W. Bruland. 1985. "Metal deposits associated with bacteria: implications for Fe and Mn biogeochemistry." *Deep-Sea Research* v. A32, n. 3, pp. 253-272.

Cowen, R. 1997. "Does the cosmos have a direction?." *Science News* v. 151, p. 252.

Cox, K. G. 1978. "Kimberlite pipes." *Scientific American* v. 238, pp. 120-132.

Cuénot, L. C. M. J. 1894a. "La Nouvelle Théorie transformiste." *Revue générale des sciences*.

———. 1894b. *L'influence du milieu sur les animaux*. Paris: Masson.

———. 1925. *L'adaptation*. Paris: G. Doin.

Cullen, J. J. 1982. "The deep chlorophyll maximum, comparing vertical profiles of chlorophyll a." *Canadian Journal of Fisheries and Aquatic Science* v. 39, pp. 791-803.

Cuvier, G. 1826. *Discours sur les révolucions de la surface du globe, et sur les changements qu'elles ont produits dans le règne animal*. Paris: G. Dufour and E. D'Ocagne.

d'Azara, F. 1905. *La Geografia Fisica y Esférica del Paraguay y Misiones Guaranies*. Taller de Impresiones Oficiales: La Plata.

Daly, M. J., and K. W. Minton. 1995. "Interchromosomal recombination in bacterium *Deinoccus radiodurans*." *Journal of Bacteriology* v. 177, pp. 5495-5505.

———. 1996. "An alternative pathway of recombination of chromosomal fragments precedes recA-dependent recombination in the radioresistant bacterium *Deinococcus radiodurans*." *Journal of Bacteriology* v. 178, pp. 4461-4471.

Daly, R. A. 1928. "The outer shells of the earth." *American Journal of Science* v. 15, pp. 110-135.

———. 1938. *Architecture of the Earth*. New York: D. Appleton-Century Company.

———. 1940. *Strength and Structure of the Earth*. New York: Prentice-Hall.

Dalziel, I. W. D. 1997. "Neoproterozoic-Paleozoic geography and tectonics: Review, hypothesis, environmental speculation." *Geological Society of America Bulletin* v. 109, pp. 16-42.

Dangeard, P. A. C. 1910a. "Les spectrogrammes en physiologie vegetale." *Paris Bull. Soc. Bot.* v. 57, pp. 91-93.

———. 1910b. "L'action de la lumière sur la chlorophylle." *Comptes Rendu de Académie des Sciences, Paris* v. 151, pp. 1386-1388.

———. 1911a. "Sur la détermination des rayous aetifs dans la synthese chlorophycesss." *Comptes Rendu de Académie des Sciences, Paris* v. 152, pp. 277-279.

———. 1911b. "Sur les conditions de l'assimilation chlorophylliene chez les cyanophycees." *Comptes Rendu de Académie des Sciences, Paris* v. 152, pp. 967-969.

———. 1911c. "Sur l'adaptation chromatique complementaire chez les végétaux." *Comptes Rendu de Académie des Sciences, Paris* v. 153, pp. 293-294.

———. 1911d. "L'action de la lumière sur la chlorophylle." *Rev. gen. agron. louvain*, pp. 51-52.

Darwin, C. 1963 [first published 1859]. *The Origin of Species by Means of Natural Selection or the Preservation of Favored Races in the Struggle for Life.* New York: Washington Square Press.

Daum, E., and W. Schenk. 1974. *A Dictionary of Russian Verbs : Bases of Inflection, Aspects, Regimen, Stressing, Meanings.* New York: Hippocrene Books.

Delaney, J. R., J. A. Baross, M. D. Lilley, D. S. Kelley, and R. W. Embley. 1994. "Is the quantum event of crustal accretion a window into a deep hot biosphere?" *Eos* v. 75, pp. 617-618.

Dendibh, K. G. 1975. *An Inventing Universe*. London: Hutchinson.

Depew, D., and B. W. Weber. 1995. *Darwinism Evolving: Systems Dynamica and the Genealogy of Natural Selection.* Cambridge, MA: MIT Press.

Deriugin, K. M. 1925. "Reliktovoe Ozero Mogilnoe (ostrov Kildin v Barentsovom more)" ["Relict Lake Mogilnoe (Kildin Island, Barents Sea)"]. *Trudy Petergov. estestv.-nauchn. inst.* [*Proceedings of the Petergof. natural-research institute*] n. 2, pp. 1-112.

Desroche, P. 1911a. "Action de diverses radiations lumineuses sur le mouvement des zoospores de *Chlamydomonas*." *Comptes Rendu de Académie des Sciences, Paris* v. 153, pp. 829-832.

———. 1911b. "Mode d'action des lumières colorées sur les *Chlamydomonas*." *Comptes Rendu de Académie des Sciences, Paris* v. 153, pp. 1014-1017.

———. 1911c. "Sur l'action des diverses radiations lumineuses sur les *Chlamydomonas*." *Paris C. R. Ass. Franc. Avanc. Sci. Dijon* v. 40, pp. 485-487.

———. 1911d. "Sur le phototropisme des zoospores de *Chlamydomonas steinii*." *Comptes Rendu de Académie des Sciences, Paris* v. 152, pp. 890-893.

———. 1911e. "Sur une interpretation de la loi de Weber-Fechner." *Paris C. R. Soc. Biol.* v. 70, pp. 571-573.

Dijksterhuis, E. J. 1961. *The Mechanization of the World Picture*. Oxford: Clarendon Press.

Dobrovolsky, V. V. 1994. *Biogeochemistry of the World's Land*. Boca Raton, Florida: CRC Press.

Dover, C. L. V. 1996. *The Octopus's Garden: Hydrothermal Vents and Other Mysteries of the Deep Sea*. Reading, Massachusetts: Addison-Wesley.

Driesch, H. 1914. *The History and Theory of Vitalism*. London: Macmillan and Company.

Ducloux, E. 1905. "Sur une piroplasmose bacilliforme de boeuf en tunisie." *Paris C. R. Soc. Biol.* 59, pp. 461-463.

———. 1910. "Sur un protozoaire dans la lymphangite epizootique du mulet en Tunisie." *Paris C. R. Soc. Biol.* 64, pp. 593-595.

Dumas, J. B., and J. B. Boussingault. 1844. *Essai de statique chimique des êtres organisés*. Paris: Fortin et Masson.

Durrani, S. A. 1975. "Nuclear reactor in the jungle." *Nature* v. 256, p. 264.

Duve, C. De. 1995. *Vital Dust: Life as a Cosmic Imperative*. New York: Basic Books.

Eddington, A. S. 1914. *Stellar Movements and the Structure of the Universe*. London: Macmillan and Company.

———. 1920. *Report on the Relativity Theory of Gravitation* [2nd Ed.]. London: Fleetway Press.

———. 1921. *Space, Time and Gravitation; An Outline of the General Relativity Theory*. Cambridge: Cambridge University Press.

———. 1922. *The Theory of Relativity, and Its Influence on Scientific Thought*. Oxford: The Clarendon Press.

———. 1923. *The Mathematical Theory of Relativity*. Cambridge: Cambridge University Press.

Egounov, M. A. 1897. "Schwefeleisen und Eisenoxydhydrat in den Böden der Limane und des Schwarzen Meeres." *Annuaire Géologique et Minéralogique de la Russie* v. 2, pp. 157-180.

Ehrenberg, C. G. 1854. *Mikrogeologie: das Erden und Felsen schaffende Wirken des unsichtbar kleinen selbstandigen Lebens auf der Erde*. Leipzig: L. Voss.

Eliseev, E. N., and I. I. Shafranovskii. 1989. "Vliyanie idei V. I. Vernadskogo na razvitie kristallografii (istoriya i sovremennost')" ["The influence of Vernadsky on the historical and contemporary development of crystallography"]. In *Nauchnoe i sotsial' noe znachenie deyatel' nosti V. I. Vernadskogo* [*Scientific and Social Significance of V. I. Vernadsky's Activity*], Yanshin, A. L., ed., pp.175-184. Leningrad: Nauka Leningrad Branch.

Engelmann, T. W. 1984. *Th. W. Engelmann, professor of physiology, Utrecht (1889-1897): some papers and his bibliography, with an introduction*. Amsterdam, The Netherlands: Rodopi.

Engelmann, W. 1861. *Bibliotheca zoologica [I] Verzeichniss der schriften uber zoologie, welche in den periodischen werken enthalten und vom Jahre 1846-1860 selbstandig erschienen sind. Mit einschluss der allgemein-naturgeschichtlichen, periodischen und paleontologischen schriften*. Bearb. von J. Victor Carus und Wilhelm Engelmann. Leipzig: W. Engelmann.

Epstein, A. G., J. B. Epstein, and L. D. Harris. 1977. "Conodont color alteration—an index to organic metamorphism." *United States Geological Survey Professional Paper 995*.

Ehrlich, P. R., and E. O. Wilson. 1991. "Biodiversity studies: science and policy." *Science* v. 253, pp. 758-762.

Fabry, C., and H. Buisson. 1913. "L' absorption de l' ultraviolett par l' ozone et la limite spectre soliare" ["Absorption of ultra-violet light by ozone and limit of solar spectrum"]. *Journ. de astrophysique* [*Astrophysical Journal*] v. 3, ser. 5, pp. 196-206. March.

Farago, C. 1996. *Leonardo Da Vinci, Codex Leicester, A Masterpiece of Science*. New York: American Museum of Natural History.

Farrington, O. C. 1901. "The constituents of meteorites." *Journal of Geology* v. 9, pp. 393-408, 522-532.

Fersman, A. E. 1933, 1934, 1937 and 1939. *Geokhimiya, T. I-IV* [*Geochemistry, volumes 1-4*]. Leningrad: ONTI-Khimteoret.

———. 1945. "The life path of Academician Vladimir Ivanovich Vernadsky (1863-1945)." *Soc. Russe. Miner., Mem.* v. 75, pp. 5-24.

———. 1946. *Rol' periodicheskogo zakona Mendeleeva v sovremennoi nauke* [*The Role of the Periodical Laws of Mendeleev in Contemporary Science*]. Moscow: Gosudarstvennoe nauchno-tekhnicheskoe izdatel'stvo khimicheskoi literatury.

———. 1958. *Geoquímica Recréativa*. [English *Geochemistry for Everyone*; French *La Géochimie Recreative*]. Moscow: Foreign Languages Publishing House.

Fesenkov, V. G. 1976. *Izbrannye trudy: solntse i solnechnaia sistema* [*Selected Works: The Sun and the Solar System*]. Moscow: Nauka.

Feuer, L. 1974. *Einstien and the Generation of Science*. New York: Basic Books.

Fischer, A. 1900. *The Structure and Functions of Bacteria* [Translated by A. Coppen Jones]. Oxford: Clarendon Press.

Fleming, A. 1908. "Some observations on the opsonic index with special references to the accuracy of the method, and to some of the sources of error." *Practicioner* v. 80, pp. 697-634.

Fossing, H., V. A. Gallardo, B. B. Jørgensen, M. Hüttel, L. P. Nielsen, H. Schulz, D. E. Canfield, S. Forster, R. N. Glud, J. K. Gundersen, J. Küver, N. B. Ramsing, A. Teske, B. Thamdrup, and O. Ulloa. 1995. "Concentration and transport of nitrate by the mat-forming sulfur bacterium *Thioploca*." *Nature* v. 374, pp. 713-715.

Fox, S. W. ed. 1965. *The Origins of Prebiological Systems and of their Molecular Matrices*. New York: Academic Press.

Fredricson, J. K., and T. C. Onstatt. 1996. "Microbes deep inside the earth." *Scientific American* v. 275, pp. 68-73.

Gal'perin, I. R. ed. 1972. *Bol'shoi Anglo-Russkii Slovar'* [*New English-Russian Dictionary*]. Moscow: Soviet Encyclopedia Publishing House.

Gallucci, V. F. 1973. "On the principles of thermodynamics in ecology." *Annual Review of Ecology and Systematics* v. 4, pp. 329-357.

Gardner, M. 1963. "Mathematical games." *Scientific American* v. 208, pp. 148-156.

Gates, D. 1962. *Energy Exchange in the Biosphere*. New York: Harper & Row.

Georgescu-Roegen, N. 1971. *The Entropy Law and the Economic Process*. Cambridge, MA: Harvard University Press.

———. 1995. *La Décroissance: entropie-écologie-économie*, présentation et traduction de Jacques Grinevald et Ivo Rens. Paris: Editions Sang de la Terre.

Germain, L. 1924. *La vie des animaux a la surface des continents*. Paris: F. Alcan.

———. 1925. *La faune des lacs, des étangs et des marais*. Paris: P. Lechevalier.

Germanov, A. I., and S. G. Melkanovitskaya. 1975. "Organicheskie kisloty v gidrotermalnom obrazovanii polymetallicheskih rud i podzemnyh vodah obolochki osadochnyh porod Zemli" ["Organic-acids in hydrothermal formations of polymetallic deposits and underground waters of the Earth's sedimentary shell"]. *Doklady Akademiya Nauk SSSR* v. 225, pp. 192-195.

Ghilarov, A. M. 1995. "Vernadsky's biosphere concept: an historical perspective." *The Quarterly Review of Biology* v. 70, pp. 193-203.

Gibbs, J. W. 1902. *Elementary Principles in Statistical Mechanics, Developed with Especial Reference to the Rational Foundation of Thermodynamics.* New York: C. Scribner's Sons.

Gillispie, C. C. 1977-1980. *Dictionary of Scientific Biography.* New York: Macmillan Reference.

Ginzburg-Karagicheva, T. L. 1926. "Mikrobiologicheskie issledovaniya serno-solenykh vod Apsherona" ["Microbiological investigations of sulphuric-haline waters of Apsheron"]. *Azerbidzhan. Neftyanoe Kozyaistvo [Azerbaijan Oil Economics]* n. 6-7, pp. 30-39.

———. 1927. "Eshche o mikroflore burovykh i sernykh vod Apsherona" ["More concerning the microflora of bore hole and sulphuric waters of Apsheron"]. *Azerbidzhan. Neftyanoe Kozyaistvo [Azerbaijan Oil Economics]* n. 5, pp. 67-69.

Glansdorff, P. and I. Prigogine. 1971. *Thermodynamic Theory of Structure, Stability and Fluctuation.* New York: Wiley-Interscience.

Glanz, J. 1997. "Doubts greet claim of cosmic axis." *Science* v. 276, p. 530.

Glick, T. F., ed. 1974. *The Comparative Reception of Darwinsim.* Austin, Texas: University of Texas Press.

Gold, T. 1996. "Outward bound." *Newsweek* October 14, v. 128, p. 18.

Gold, T., and S. Soter. 1980. "The deep-earth-gas hypothesis." *Scientific American* v. 242, pp. 154-165.

Goldschmidt, V. M. 1929. "The distribution of the chemical elements." *Proceedings of the Royal Institution* v. 26, pp. 73-86. Condensed in *Nature* v. 124, n. 3114, pp. 15-17.

———. 1954. *Geochemistry.* Muir, A., ed. Oxford: Clarendon Press.

Gould, J. L., and C. G. Gould, eds. 1989. *Life at the Edge: Readings from Scientific American.* New York: Freeman.

Graham, L. 1993. *Science in Russia and the Soviet Union. A Short History.* New York: Cambridge University Press.

Greene, M. T. 1982. *Geology in the Nineteenth Century.* Ithaca: Cornell University Press.

Greenwood, D. 1964. *Mapping.* Chicago: The University of Chicago Press.

Grinevald, J. 1987. *Vernadsky and Lotka as a Source for Georgescu-Roegen's Bioeconomics.* Vienna Centre International Conference on Economics and Ecology, 26-29 September. Spain: Universitat Autonoma de Barcelona [Spanish version in *Ecologia Politica*, 1990 v. 1, pp. 99-112].

———. 1987. "On a holistic concept for deep and global ecology: The Bioshere." *Fundamenta Scientiae* v. 8, n. 2, pp. 197-226.

———. 1988. "Sketch for a History of the Idea of the Biosphere." In *Gaia: The Thesis, the Mechanisms, and the Implications,* Bunyard, P., and E. Goldsmith, eds., pp. 1-34. Cornwall: Wadebridge Ecological Centre.

———. 1990. *The Industrial Revolution and the Earth's Biosphere: A Scientific Awareness in Historical Perspective. Selected bibliographical notes.* ProClim PWS. 4, Publ. 90/1. Bern: Swiss Academy of Sciences.

———. 1996. "Biodiversité et Biosphère." *L'Etat de la Planète-Magazine* November/December, pp. 8-11.

Gross, M. G. 1982. *Oceanography, A View of the Earth* [3rd ed.]. Englewood Cliffs, New Jersey: Prentice Hall.

Gruner, J. W. 1924. *Contributions to the Geology of the Mesabi Range: with Special Reference to the Magnetites of the Iron-bearing Formation West of Mesaba.* Minneapolis: University of Minnesota Press.

———. 1946. *The Mineralogy and Geology of the Taconites and Iron Ores of the Mesabi Ranges.* Minnesota, St. Paul: Minnesota, Office of the Commissioner of the Iron Range Resources and Rehabilitation.

Guegamian, G. V. 1980. "O biosferologii Vernadskogo" ["On Vernadsky's Bios-
pherology]." *Zhurnal Obshchei Biologii* v. 41, pp. 581-595.

Gupta, S. M. 1987. "Paleogene ichthyoliths from the substrates of ferromanganese encrustations and nuclei of the manganese nodules from the central Indian Ocean Basin." *Journal of the Palaeontological Society of India* v.
32, pp. 85-91.

Gurevich, M. S. 1964. *Rol' mikroorganizmov v obrazovanii zhelezo-margantsevykh ozernykh rud.* [*The Role of Microorganisms in the Development of
Iron-Manganese Lake Deposits*]. Leningrad: Nauka.

Gutenberg, B. 1924. "Der Aufbau der Erdkruste auf Grund geophysikalischer
Betrachtungen." *Z. Geophys.* Bd. 1, pp. 94-108.

———. 1925. *Der Aufbau der Erde.* Berlin: Gebr. Borntraeger.

Guterl, F. 1997. "An Odyssey of sorts." *Discover* v. 18, pp. 68-69.

Haldane, J. B. S. 1924. *Daedalus; or, Science and the Future.* London: K. Paul,
Trench, Trubner and Company.

Hallam, A. 1992. *Great Geological Controversies* [2nd ed.]. Oxford: Oxford Science Publications.

Hann, J. 1883. *Handbuch der Klimatologie.* Stuttgart, Germany: J. Engelhorn.

Hardy, A. C. 1953. "On the origin of the Metazoa." *Quarterly Journal of Micro-
scopical Science* v. 94, pp. 441-443.

Harkins, W. D. 1917. "The evolution of the elements." *Journal of the American
Chemical Society* v. 39, pp. 859-879.

Harris, A. G. 1979. "Conodont color alteration, an organo-mineral metamorphic index, and its application to Appalachian Basin geology." *Society of
Economic Paleontologists and Mineralogists, Special Publication* n. 26,
pp. 3-16.

Harris, A. G., L. D. Harris, and J. B. Epstein. 1978. "Oil and gas data from Paleozoic rocks in the Appalachian basin: Maps for assessing hydrocarbon
potential and thermal maturity (conodont color alteration isograds and
overburden isopachs)." *United States Geological Survey Miscellaneous
Investigation Series Map* v. 1, scale 1: 2,500,000.

Hayes, J. M. 1996. "The earliest memories of life on Earth." *Nature* v. 384, pp.
21-22.

Henderson, L. J. 1913. *The Fitness of the Environment.* New York: Macmillan &
Co.

Hess, V. F. 1928. *The electrical conductivity of the atmosphere and its causes.*
New York: D. Van Nostrand Company.

Hesselman, H. 1917. *Studier over salpeterbildningen i naturliga jordmaner
och dess betydelse i vaxtekologiskt avseende. Studien über die nitratbil-
dung in naturkichen boden und ihre bedeutung in pflanzenekoloischer
hinsicht.* Stockholm: Centraltryckeriet.

Hingston, R. W. G. 1925. "Animal life at high altitudes." *Geographical Journal*
v. 65, n. 3, pp. 185-198. March.

Hinkle, G. J. 1996. "Marine salinity: Gaian phenomenon?" In *Gaia in Action:
Science of the Living Earth*, Bunyard, P., ed., pp. 99-104. Edinburgh, Scotland: Floris Books.

Hitchcock, D. R., and J. E. Lovelock. 1967. "Life detection by atmospheric
analysis." *Icarus* v. 7, pp. 149-159.

Hjort, J., and H. H. Gran. 1900. *Hydrographic-biological investigations of the
Skagerrak and the Christiana Fiord.* Kristiania, Norway: Oscar Anderson.

Ho Ahn, J., and P. R. Busek. 1990. "Hematite nanospheres of possible colloidal origin from a Precambrian Banded Iron Formation." *Science* v. 250,
pp. 111-113.

Hoffmann, C. 1949. "Über die durchlässigkeit dünner sandschichten für licht"
["On the penetration of light through thin layers of sand"]. *Planta* v. 36, pp.
48-56.

Holland, H. D. 1990. "The origins of breathable air." *Nature* v. 347, p. 17.

Howard-Bury, C. K. 1922. "The Mount Everest Expedition." *Geographical Journal* v. LIX, n. 2, pp. 81-99.

Huggett, R. J. 1991. *Climate, Earth Processes and Earth History*. Heidelberg: Springer-Verlag.

———. 1995. *Geoecology: An Evolutionary Approach*. London, New York: Routledge.

Humbolt, A. von. 1859. *The life, travels, and researches of Baron Humboldt*. London: Nelson.

Hutchinson, G. E. 1954. "The Biochemistry of the Terrestrial Atmosphere." In *The Earth as a Planet: The Solar System*. Kuiper, G. P., ed., v. 2, pp. 371-433. Chicago: University of Chicago Press.

———. 1957-1992. *Treatise of Limnology* vols. I-IV. New York: John Wiley & Sons.

———. 1964. "Notes on ecological principles easily seen in the hall of Southern New England Natural History." *Yale Peabody Museum Special Publication* v. 8, p. 7.

———. 1965. *The Ecological Theater and the Evolutionary Play*. New Haven: Yale University Press.

———. 1970. "The Biosphere." *Scientific American* v. 223, n. 3, pp. 45-53 [In *The Biosphere*. A Scientific American Book. San Francisco: Freeman. pp. 3-11].

Hutton, J. 1795. *Theory of the Earth, With Proofs and Illustrations* [2 vols]. Edinburgh: William Creech [Facsimile reprint in 1959 by Wheldon and Wesley, Codicote, Herts.].

Huxley, A. 1921. *Chrome Yellow*. London: Chatto and Widus.

Isachenko, B. L. 1927. "Mikrobiologicheskie issledovania nad gryazevymi ozerami" ["Microbiological studies of mud lakes"]. *Trudy Geologich. kom.* [*Transactions of the Geological Committee*] v. 148, pp. 1-154.

Janin, M. C. 1987. "Micropaleontology of manganese nodules from the Equatorial North Pacific Ocean; Area SO 25-1 and SO 25-3." *Geologisches Jahrbuch. Reihe D. Mineralogie, Petrographie, Geochemie, Lagerstaettenkunde* v. 87, pp. 315-375.

Jeffreys, H. 1924. *The Earth: Its Origin, History and Physical Constitution*. Cambridge: Cambridge University Press.

Jerzmanska, A., and J. Kotlarczyk. 1976. "The beginning of the Sargasso assemblage in the Tethys." *Palaeogeography, Palaeoclimatology, Palaeoecology* v. 20, pp. 297-306.

Johnstone, J. 1908. *Conditions of Life in the Sea: A Short Account of Quantitative Marine Biological Research*. Cambridge: Cambridge University Press.

———. 1911. *Life in the Sea*. Cambridge: Cambridge University Press.

———. 1926. *A Study of the Oceans*. London: Edward Arnold.

Judson, H. F. 1979. *The Eighth Day of Creation*. New York: Simon and Schuster.

Kalugin, A. S. 1970. *Atlas tekstur i struktur vulkanogenno-osadochnykh zheleznykh rud Altaya* [*Atlas of Structures and Textures of Volcanogenic-sedimentary Iron Ores in the Altai*]. Leningrad: Izdatel'stvo "Nedra".

Kamshilov. M. M. 1976. *Evolution of the Biosphere* [Translated by M. Brodskaya]. Moscow: MIR Publishers.

Kant, I. 1981 [first published 1755]. *Universal natural history and theory of the heavens*. Edinburgh: Scottish Academic Press.

Karavaiko, G. N., S. I. Kuznetsov, and A. I Golomzik. 1972. *Rol' mikroorganismov v byshchelachivanii metallov iz rud* [*The role of microorganisms in the leaching of metals from ore*]. Moscow: Izdatel'stvo Nauka.

Karol', I. L., A. A. Kiselev, and V. A. Frol' kis. 1995. "Is it possible to 'repair' the ozone holes?" *Izvestia–Atmospheric and Oceanic Physics* [In English] v.

31/1, pp.113-115.

Kelvin, Lord. 1894. *Popular Lectures and Addresses*. London: Macmillan & Co.

Kerr, R. A. 1997. "Life goes to extremes in the deep Earth—and elsewhere?" *Science* v. 276, pp. 703-704.

Khakhina, L. N. 1988. "Razvitie B. M. Kozo-Polyanskim problemy faktorov makroevolyutsii." In *Darvinizm: istoriya i sovremennost'* [*Historical and Modern Darwinism*], Kolchinskii, E. I., and Yu. I. Polyanskii, eds., pp. 178-183. Leningrad: Nauka.

Kholodov, V. N., and R. K. Paul'. 1993. "Problems of ancient phosphorite genesis." *Lithology and Mineral Resources* n. 3, pp. 110-125.

Kim-Tekhn. Izdatel'stvo. 1960. *Izbrannye Sochineniya* [*Selected Works*]. Moscow: Izdatel'stvo Akademiya Nauk SSSR v. 5, pp. 7-102.

Klossovsky, A. V. 1908. *Meteorologia. Obshchy Kurs* [*Meteorology. General Handbook*], Part 1. Odessa: Ekonomich. Tipogr.

———. 1914. *Osnovy Meteorologii* [*Principles of Meteorology*]. Odessa: Mathesis.

Kolchinsky, E. I. 1985. "On the main tendencies in the evolution of biosphere." In *Evolution and Morphogenesis*, Milkovsky, J., and V. J. A. Novak, eds., p. 771-778. Prague: Academia Press.

———. 1987. *Idei V. I. Vernadskogo ob evolyutsii biosfery (k 125-letiyu co dnya rozhdeniya)* [*The ideas of V. I. Vernadsky concerning the evolution of the biosphere (on the 125th anniversary of his birth)*]. Leningrad: Leningradskaya organizatsiya, Obshchestvo "Znanie," RSFSR.

———. ed. 1988. *V. I. Vernadskii i sovremennaya nauka* [*V. I. Vernadsky and Contemporary Science*]. Leningrad: Nauka Leningradskoe Otdelenie.

Kolchinsky, E. I., and S. A. Orlov. 1990. *Filosofiskie problemy biologii v SSSR (1920—1960 gg)* [*Philosophical Problems of Biology in the USSR, 1920-1960*]. Leningrad: Akademiya Nauk SSSR, Leningradskii Otdel Instituta Istorii Estestvoznaniya i Tekhniki, Mezhdunarodnyi Fond Istorii Nauka.

Konhauser, K. O., and F. G. Ferris. 1996. "Diversity of iron and silica precipitation by microbial mats in hydrothermal waters, Iceland: Implications for Precambrian iron formations." *Geology* v. 24, pp. 323-326.

Kovda, V. A. 1985. *Biogeokhimia Pochvennogo Pokrova* [*Biogeochemistry of the Soil Mantle*]. Moscow: Nauka.

Kramarenko, L. E. 1983. *Geokhimicheskoe i poiskovoe znachenie mikroorganizmov podzemnykh vod* [*Geochemical Importance and Prospecting Potential of Groundwater Microorganisms*]. Leningrad: Nedra.

Krishnan, R., et. al., eds. 1995. *A Survey of Ecological Economics*. Washington, D. C.: Island Press.

Kropotkin, P. A. 1902. *Mutual Aid: A Factor in Evolution*. Facsimile reprint in 1987 by Freedom Press, London.

Krumbien, W. E., D. M. Paterson, and L. J. Stal, eds. 1994. *Biostabilization of Sediments*. Oldenburg, Germany: Bibliotheks- und Informations system der Universitaet Oldenburg.

Kuhn, T. S. 1962. *The Structure of Scientific Revolutions*. Chicago: University of Chicago Press.

Kump, L. 1993. "Bacteria forge a new link." *Nature* v. 362, pp. 790-791.

Kurkin, K. A. 1989. "Sistemichno-paragenemichniy metod izucheniya biosfernoi evolutsii u printsipa deistvitel'nosti" ["Systematic-paragenetic method of study of biospheric evolution and the principle of actualism"]. *Zhurnal Obshchei Biologii* v. 50, pp. 516-528.

Kuznetsov, P. G. 1965. "Concerning the hitosry of the question of applying thermodynamics to biology." In *Biology and Information: Elements of Biological Thermodynamics*, traduit du russe par E. S. Spiegelthal, Trincher, K. S., pp. 83-93. New York: Consultants Bureau.

Kuznetsov, S. I., M. V. Ivanov, and N. N. Lyalikova. 1962. *Vvedenie v geo-*

logicheskuyu mikrobiologiyu [Introduction to geological microbiology].
Moscow: Izdatel'stvo Akademiya Nauk SSSR.

Lamarck, J. B. 1964 [first published 1802]. Hydrogeology [translated by A. V.
Carozzi]. Urbana, Illinois: University of Illinois Press.

Langly, L. L., ed. 1973 Homeostasis: Origins of the concept, Benchmark
Papers in Human Physiology. Stroudsburg, Dowden: Hutchinson and Ross.

Lapo, A. V. 1979. Sledy Bylykh Biosfer. Moscow: Lzdatel'stvo Znanie.

————. 1980. "Problemy biogeokhimii" ["Problems of Biogeochemistry"].
Trudy BIOGEL (Trans. Biogeochem. Lab.) v. 16, pp. 1-320.

————. 1987. Traces of Bygone Biospheres [translated by V. Purto]. Oracle,
Arizona: Synergetic Press, and Moscow: MIR Publishers.

————. 1988. "Vernadskii–Lomonosov XX veka" ["Vernadsky–A Lomonosov
of the 20th Century"]. In Byulleten' komissii po razraboke nauchnogo
naslediya akademika V. I. Vernadskogo No. 4 [Bulletin of the Commission
for the Dissemination of the Scientific Heritage of Academician V. I. Vernadsky, No. 4], Melya, A. I., ed., pp. 3-10. Leningrad: Adademiya Nauk
SSSR.

————. 1989. "Rol' zhivykh organizmov v protsessakh litogeneza" ["The role
of living organisms in the process of lithogenesis"]. In Teoretich. i priklad-
nye aspekty sovrem. paleontologii [Theoretical and applied aspects of
contemporary paleontology], pp. 13-24. Leningrad: Nauka.

————. 1990. "Problema geologicheskoi deyatel'nosti mikroorganismov v
trudakh V. I. Vernadskogo" ["The problem of the geological activity of
microorganisms in the work of V. I. Vernadsky"]. In Problem' sovremennoi
paleontologii, 34th session VPOI, pp. 201-208. Leningrad: Nauka.

Lavoiser, L. A. 1892. La Chaleur et la Respiration. Paris: Masson.

Leibig, F. V. J. 1847. Organische Chemie in Ihrer Anwendung Auf Agricultur and
Physiologie [Chemistry in its Applications to Agriculture and Physiology].
London: Taylor and Walton.

Lenin, V. I. 1909. Materialism and Empirio-Criticism. New York: International
Publishers.

Leontiev, P. 1927 [Leont'ev, I. F.]. "Specific Gravity of Protoplasm." Jour. Bio. et
Med. Exp. v. 5(15), pp. 83-89.

Lepsky, S. D. 1980. "O biogeneticheskih aspektah kontzentrirovanija urana
pri sedimentatzii" ["On biogenic aspects of uranium concentrations under
recent sedimentation"]. Dopovidi Akademii Nauk Ukrainskoi RSR Seriya
B—Geologichni Khimichni ta Biologichni Nauki v. 8, pp. 15-18.

Levi, P. 1995. "The story of a carbon atom." In The Faber Book of Science,
Carey, J., ed., pp. 338-344. Wincester, Massachusetts: Faber and Faber.

Levine, J. S. 1985. The Photochemistry of Atmospheres—Earth, the Other
Planets, and Comets. Orlando, Florida: Academic Press.

Li, Z. X., I. Metcalfe, and C. M. Powell. 1996. "Breakup of Rodinia and Gond-
wanaland and assembly of Asia." Australian Journal of Earth Sciences v.
43, pp. 591-699.

Lieth, H., and R. H. Whittaker, eds. 1975. Primary Productivity of the Biosphere. New York: Springer-Verlag.

Lindsay, R. B. 1973. Julius Robert Mayer: Prophet in Energy. Oxford: Pergamon Press.

Linnaeus, C. 1964 [first published 1759]. Carl Linnaeus Systema nature: t. II :
Vegetabilia. New York: Stechert-Hafner.

Lisitsyn, A. P. 1978. Protsessy okeanskoi sedimentatsii [Processes of Oceanic
Sedimentation]. Moscow: Nauka.

Losos, J. B., K. I. Warheit, and T. W. Schoener. 1997. "Adaptive differentiation
following experimental island colonization in Anolis lizards." Nature v. 387,
pp. 70-73.

Lotka, A. J. 1925. Elements of Physical Biology. Baltimore, Maryland: Williams
and Wilkins Company.

————. 1945. "The law of evolution as a maximal principle." *Human Biology* v. 17, pp. 167-194.

Lovelock, J. 1979. *Gaia: A New Look at Life on Earth*. Oxford: Oxford University Press.

————. 1983. "Daisy World: A Cybernetic Proof of the Gaia Hypothesis." *The CoEvolution Quarterly* n. 38, pp. 66-72.

————. 1986. "The Biosphere." *New Scientist* July 17, p.51.

————. 1988. *Ages of Gaia*. New York: W.W. Norton & Company.

Lovins, A. B., L. H. Lovins, F. Krause, and W. Bach, 1981. *Least-Cost Energy— Solving the CO$_2$ Problem*. Andover, Massachusetts: Brick House Publishing Company.

Lowell, P. 1909. *The Evolution of Worlds*. New York: Macmillan & Co.

Lutz, R. A. et al. 1994. "Rapid growth at deep-sea vents." *Nature* v. 371, p. 663-664.

Lyell, C. 1830-1833. *Principles of Geology, Being an Attempt to Explain the Former Changes of the Earth's Surface by Reference to Causes Now in Operation*. London: John Murray. Facsimile edition published in 1990 by University of Chicago Press, Chicago.

Lysenko, S. V. 1979. "Microorganisms in the upper atmospheric layers." *Microbiology* v. 48, pp. 871-878.

Lysenko, T. 1948. *Agrobiologiya*, [5 ed.] Moscow: Sel'khozgiz.

MacFadyen, A. 1902. "On the influence of the prolonged action of the temperature of liquid air on mirco-organisms, and on the effect of mechanical trituration at the temperature of liquid air on photogenic bacteria." *Proc. Royal Soc.* 71, p. 76.

Malone, T. F., and J. G. Roederer, eds. 1985. *Global Change*. Cambridge: Cambridge University Press, p.xiii.

Manahan, S. E. 1994. *Environmental Chemistry* [6th ed.]. Boca Raton, Louisiana: Lewis Publishers.

Mann, H., and W. S. Fyfe. 1985. "Algal uptake of U and some other metals: Implications for global geochemical cycling." *Precambrian Research* v. 30, pp. 337-349.

Margulis, L., and D. Sagan. 1985. "Biospheric concepts—the real deficit: our debt to the biosphere," In *The Biosphere Catalog*, T. P. Snyder, ed., pp. 1-3. London, Great Britain and Fort Worth, Texas: Synergetic Press.

————. 1995. *What Is Life?*. New York: Nevraumont/Simon and Schuster.

————. 1997. *Slanted Truths: Essays on Gaia, Symbiosis and Evolution*. New York: Copernicus.

Mason, B. 1950. *The Principles of Geochemistry* [4th ed., 1982]. New York: John Wiley & Sons.

————. 1954. "The geochemistry of the crust." In *The Earth as a Planet* ("The Solar System"), G. P. Kuiper, ed., pp. 258-298. Chicago: University of Chicago Press.

————. 1992. *Victor Moritz Goldschmidt: Father of Modern Geochemistry*. San Antonio, Texas: Geochemistry Society.

Mason B., and C. B. Moore. 1982. *The Principles of Geochemistry* [4th ed.]. New York: John Wiley & Sons.

Masterton, W. L., and E. J. Slowinski. 1966. *Chemical Principles*. Philadelphia and London: W. B. Saunders Company.

Matias, M. J. S., and M. G. Habberjam. 1984. "Book Review of *Vladimir Vernadsky* by R. K. Balandin." *Geophysics* v. 49, p. 1567.

Mattimore, V., and J. R. Battista. 1996. "Radioresistance of *Deinococcus radiodurans*: Functions necessary to survive ionizing radiation are also necessary to survive prolonged desiccation." *Journal of Bacteriology* v. 178, pp. 633-637.

Maturana, H. R., and F. J. Varela. 1980. *Autopoeisis and Cognition*. Dordrecht,

Holland: D. Reidel [Translation of 1972, *De Maquinas y Seres Vivos*, Editorial Universitaria S. A.].

Mayer, J. 1845. "The Motions of Organisms and their Relation to Metabolism." In *Julius Robert Mayer: Prophet of Energy*, Lindsay, R. B. 1973. Oxford: Pergamon Press.

———. 1855. "Quantitative Analysis of the Spring Water of Ramandroog." *Indian Annals* II, pp. 239-242.

McCammon, C. 1997. "Perovskite as a possible sink for ferric iron in the lower mantle." *Nature* v. 387, pp. 694-696.

McMenamin, M. 1986. "The Garden of Ediacara." *Palaios* v. 1, pp. 178-182.

———. 1993. "Osmotrophy in fossil protoctists and early animals." *Invertebrate Reproduction and Development* v. 23, pp. 165-169.

McMenamin, M. A. S. 1997. "Review of *The Molecular Biology of Gaia*, by George R. Williams, 1997." In *Historical Biology*. New York: Columbia University Press, in press.

McMenamin, M. A. S., and D. L. S. McMenamin. 1990. *The Emergence of Animals: The Cambrian Breakthrough*. New York: Columbia University Press.

———. 1993. "Hypersea and the Land Ecosystem." *BioSystems* v. 31, pp. 145-153.

———. 1994. *Hypersea: Life on Land*. New York: Columbia University Press.

McMenamin, M. A. S., and L. Margulis. 1992. "Note on translation and transliteration." In *Concepts of Symbiogenesis: A Historical and Critical Study of the Research of Russian Botanists*, Khakhina, L. N., Margulis L., and M. A. S. McMenamin, eds., p. xxix. New Haven: Yale University Press [originally published in Russian as *Problema simbiogeneza: istoriko-kritichesky ocherk issledovany otechestvennykh botanikov*. 1979. Leningrad, Nauka].

Mendeleev, D. [Mendeléeff]. 1897. *The Principles of Chemistry* [Translated by G. Kamensky]. London: Gongmans, Green and Company.

Mengel, O. 1923. "Caractère climatique de Font-Romeu et de Mont Louis." *Memorial de l'Office national Meteorologique de France* Ann. 1, n. 5, pp. 1-16.

Menzel, D. W., and J. H. Ryther. 1960. "The annual cycle of primary production in the Sargasso Sea off Bermuda." *Deep Sea Research* v. 6, pp. 351-367.

Meyer, W. B. 1996. *Human Impact on the Earth*. Cambridge: Cambridge University Press.

Mikhailovskii, V. N. 1988. "Ot Prigozhena k Vernadskomy" ["From Prigogine to Vernadsky"]. In *V. I. Vernadskii i sovremennaya nauka* [*V. I. Vernadsky and Modern Science*], Kolchinskii, E. I., ed., pp. 99-100. Leningrad: Nauka Leningradskoe Otdelenie.

Mikulinsky, S. R. 1983. "Vernadsky as a historian of science." *Scientia* v. 118, pp. 537-560.

———. 1984. "Sarton and Vernadsky." *Isis* v. 75, pp. 56-62.

Mitchell, R. H. 1986. *Kimberlites: Mineralogy, Geochemistry, and Petrology*. New York: Plenum Press.

Mitropolsky, Yu., and M. Kratko. 1988. "Between Scylla and Charybdis." *Science in the USSR* v. 3, pp. 27-43.

Mohorovičič, A. 1910. "Das Beben vom 8.x.1909." *Jahrb. Meteorol. Observ. Zagreb* Jahrg. 9, Teil IV, Abschn. 1, pp. 1-63.

Mohorovičič, S. 1915. "Die reduzierte Laufzeitkurve und die Abhangigkeit der Herdtiefe eines Bebens von der Entfernung des Inflexionspunkted der Laufzeitkurve." *Gertland's Beitr. Geophys.* Bd. 14, pp. 187-205.

Mojzsis, S. J., G. Arrhenius, K. D. McKeegan, T. M. Harrison, A. P. Nutman, and C. R. L. Friend. 1996. "Evidence for life on Earth before 3,800 million years ago." *Nature* v. 384, pp. 55-59.

Molchanov, V. I., and V. V. Pazaev. 1996. "O prirode kisloroda vozdukha v svete ideii V. I. Vernadskogo" ["On the nature of atmospheric oxygen in the light of V. I. Vernadsky"]. *Doklady Ross. Akad. nauk* [*Papers of the Russian Academy of Sciences*] v. 349, pp. 387-388.

Monastersky, R. 1997. "Deep dwellers: microbes thrive far below ground." *Science News* v. 151, p. 192.

Monod, J. 1971. *Chance and Necessity; An Essay on the Natural Philosophy of Modern Biology* [Translated from French by A. Wainhouse]. New York: Alfred A. Knopf.

Morowitz, H. J. 1968. *Energy Flow in Biology: Biological Organization as a Problem in Thermal Physics*. New York: Academic Press.

———. 1981. "A Leap of the Imagination." *Hospital Practice* v. 16, pp. 41-42.

Munn, R. E., ed. 1971-1996. *Scope* (Scientific Committee on Problems of the Environment) Vols. 1-57. New York: John Wiley & Sons.

Murray, J. 1913. *The Ocean*. New York: Henry Holt & Co.

Murray, J., and R. Irvine. 1893. "On the chemical changes which take place in the composition of sea-water associated with blue muds on the floor of the ocean." *Transactions of the Royal Society of Edinburgh* v. 37, pp. 481-507.

Nafi Toksöz, M. 1975. "The subduction of the lithosphere." *Scientific American* v. 233, pp. 89-98.

Nafi Toksöz, M., J. W. Minear, and B. R. Julian. 1971. "Temperature field and geophysical effects of a downgoing slab." *Journal of Geophysical Research* v. 76, pp. 1113-1138.

Nagy, B., F. Gauthierlafaye, P. Holliger, J. Mossman, and J. S. Leventhal. 1993. "Role of organic matter in the Proterozoic Oklo natural fission reactors, Gabon, Africa." *Geology* v. 21, p. 655-658.

Neruchev, S. G. 1982. *Uran i Zhizn' v Istorii Zemli* [*Uranium and the Life in the History of the Earth*]. Leningrad: Nedra.

Newton, I. 1989 [first published 1687]. *The preliminary manuscripts for Isaac Newton's 1687 Principia, 1684-1685 : facsimiles of the original autographs, now in Cambridge University Library*. New York: Cambridge University Press.

Nixon, P. H. ed. 1973. *Lesotho Kimberlites*. Maseru: Lesotho Development Corporation.

Nodland, B., and J. Ralston. 1997. "Indication of anisotropy in electromagnetic propagation over cosmological distances." *Physical Review Letters* v. 78, pp. 3043-3046.

Norris, H. N. 1919. "Some problems of sidereal astronomy." *Proceedings of the National Academy of Sciences (USA)* v. 5, pp. 391-416.

Ocken, H. 1843 [Oken]. *Lehrbuch der naturphilosophie*. Zurich: F. Schulthess.

Oddo, G. 1914. "Die Molekularstruktur der radioaktive Atome." *Zeitschrift. anorg. Chem.* Bdd. 63, pp. 355-380.

Odum, E. P. 1968. "Energy flows in ecosystems: a historical review." *American Zoologist* v. 8, pp. 11-18.

———. 1971a. *Environment, Power and Society*. New York: Wiley-Interscience.

———. 1971b. *Fundamentals of Ecology* [3rd ed.]. Philadelphia: W. B. Saunders.

Omeliansky, V. L. 1923. *Svyazivanie atmosfernogo azota pochvenimi mikrobami* [*Fixation of atmospheric nitrogen by soil microorgansims*]. Petrograd: Monograph published by the Commision of Studies of Natural Production, Russia Academy of Science.

Oparin, A. I. 1938. *The Origin of Life* [Translated by Sergius Morgulis]. New York: The Macmillan Company.

———. 1957. *The Origin of Life on Earth*. New York: Academic Press.

Oparin, A. I., and W. Fox, eds. 1965. *The Origins of Prebiological Systems and of their Molecular Matrices*. New York: Academic Press.

Osborn, H. F. 1917. *The Origin and Evolution of Life on the Theory of Action,*

Reaction and Interaction of Energy. New York: C. Scribner's Sons, and 1918, London: G. Bell.

Paczoski, J. K. 1908. *Prichernomorskie Stepi. Botaniko-geografichesky ocherk* [*Steppes near the Black Sea. Botanical and Geographic Essay*]. Odessa: Slavyanskaya Tipografia.

Padian, K. 1985. "The origins and aerodynamics of flight in extinct vertebrates." *Palaeontology* v. 28, pp. 413-433.

Pasteur, L. 1876. *Etudes sur la bière, ses maladies, causes qui les provoquent, procédé pour la rendre inalterable; avec une théorie nouvelle de la fermentation*. Paris: Gauthier-Villars.

Patrick, R. 1973. "Use of algae, especially diatoms, in the assessment of water quality," In *Biological Methods for the Assessment of Water Quality*, Cairns, J., and K. L. Dickson, eds., pp. 76-95. Philadelphia, Pennsylvania: American Society for Testing and Materials Publication 528.

Paul, A. Z., E. M. Thorndike, L. G. Sullivan, B. C. Heezen, and R. D. Gerard. 1978. "Observation on the deep-sea floor from 202 days of time lapse photography." *Nature* v. 272, pp. 812-814.

Payne, C. H. 1925. *Stellar Atmospheres: A Contribution to the Observational Study of High Temperature in the Reversing Layers of Stars, Harvard Observatory Monographs, N. 1*. Cambridge, Massachusetts: The Harvard Observatory.

Peak, M. J., F. T. Robb, and J. G. Peak. 1995. "Extreme resistance to thermally induced DNA backbone breaks in the hyperthermophilic archaeon *Pyrococcus furiosus*." *Journal of Bacteriology* v. 177, pp. 6316-6318.

Pearl, R. 1912. "Notes on the history of barred breeds of poultry." *Biological Bulletin* v. 22, pp.297-300.

Pedersen, K. 1993. "The deep subterranean biosphere." *Earth-Science Reviews* v. 34, pp. 243-260.

Perelman, A. I. 1977. *Biokosnye Sistemy Zemli* [*Bio-inert Systems of the Earth*]. Moscow: Nauka.

————. 1979. *Geokhimia* [*Geochemistry*]. Moscow: Vysshaya Shkola.

Perfil'ev, B. V. 1926. "Novye dannye o roli mikroorganismov v rudoobrazovanii" ["New data about the role of microbes on ore origin"]. *Izvestia Geologich. kom.* [*Proceedings of the Geological Committee*] v. 45, n. 7, pp. 795-819.

————. 1964. "Kapillyarnyi metod mikrobnogo peizazha v geomikrobiologii" ["The capillary action of the microbial lawn in geomicrobiology"]. In *S. Rol' mikroorganismov v obrazovanii zhelezo-margantsevykh ozernykh rud* [*The role of microorganisms in the formation of iron-manganese lacustrine ores*]. Gurevich, M., ed., pp. 6-15. Leningrad: Nauka, Laboratoriya gidrogeologicheskikh problem, Akademiya Nauk SSSR.

Pestana, H. 1985. "Carbonate sediment production by Sargassum epibionts." *Journal of Sedimentary Petrology* v. 55, pp. 184-186.

Pfeffer, W. 1881. *Pflanzenphysiologie: ein Handbuch des Stoffwechels und Kraftwechels in der Pflanze, Bd. 1 & 2*. Leipzig: Engelmann.

Poirier, J.-P. 1997. "Aluminum under the spotlight." *Nature* v. 387, pp. 653-654.

Polunin, N. 1972. "The Biosphere Today." In *The Environmental Future*. Proceedings of the first International Conference on Environmental Future, held in Finland from 27 June to 3 July, 1971. N. Polunin, ed., 33-52. London: Macmillan; New York: Barnes & Noble.

————. 1980. "Editorial: Environmental Education and the Biosphere." *Environmental Conservationaism* v. 7, n. 2, pp. 89-90.

————. 1984. "Genesis and Progress of the World Campaign and Council for the Biosphere." *Environmental Conservationism* v. 11, n. 4, p. 293-298.

Polunin, N. and J. Grinevald. 1988. "Vernadsky and Biospheral Ecology." Environmental Conservationism v.15, n. 2, pp. 117-122.

Pompeckj, J. F. 1928. "Is the Earth growing old?" *Smithsonian Institution, Annual Report* 1927, Publication 2936, pp. 255-270.

Povarennykh, A. S. 1970. "O znachenii opredeleniya ponyatii i terminologii dla razvitiya nauki (no primerakh mineralogii)" ["Concerning the significance of definitive knowledge and terminology for the development of science (mineralogy, for example)"]. In *Dialektika razvitiya i teoriya poznaniya v geologii* [*Dialectical development and theoretical knowledge in geology*], Povarennykh, A. S., ed., pp. 5-30. Kiev: Naukova dumka.

Press, F., and R. Siever. 1982. *Earth* [3rd ed.]. San Francisco: W. H. Freeman.

Progogine, I., and I. Stengers. 1988. *Entre le temps et l'éternité*. Paris: Libraire Arthème Fayard.

Rankama, K., and T. G. Sahama. 1950. *Geochemistry*. Chicago: University of Chicago Press.

Raup, D. M., and J. W. Valentine. 1983. "Multiple origins of life." *Proceedings of the National Academy of Sciences [USA]* v. 80, pp. 2981-2984.

Reinke, J. 1901. *Einleitung in die theoretische Biologie*. Berlin, Pactel.

Reynolds, J. H. 1960. "Determination of the age of the elements." *Phys. Rev. Letters* v. 4, pp. 8-10.

Rodin, L. E., and N. I. Basilevich. 1965. *The Dynamics of Organic Matter and Biological Rotation in Basic Types of Plants*. Moscow: Nauka.

Rogal', I. G. 1989. "V. I. Vernadskii o bosniknovenii i evolyutsii zemnoi zhisni i sovremennoe sostoyanie problemy" ["Vernadsky on the origin and evolution of life on Earth and the present state of the problem]". In *Nauchnoe i sotsial'noe znachenie deyatel'nosti V. I. Vernadskogo* [*Scientific and social significance of V. I. Vernadsky's activity*], A. L. Yanshina, ed., pp. 87-97. Leningrad: Nauka.

Romankevich, E. A. 1984. *Geochemistry of Organic Matter in the Ocean*. Berlin: Springer-Verlag.

———. 1988. "Living matter of the Earth." *Geochemistry* n. 2, pp. 292-306 [English translation].

Rowland, S. M. 1993. "Review of *Science and Russian Culture in an Age of* Revolution by Kendall E. Bailes." *Earth Sciences History* v. 12, pp. 245-247.

Ruse, M. 1988. *Philosophy of Biology Today*. Albany, New York: State University of New York Press.

Saha-Meg-Nad. 1927. "Über ein neues Schema für den Atomaufbau." *Physikalische Zeiteitschrift* Bd. 28, pp. 469-479.

Saukov, A. A. 1950. *Geokhimiya* [*Geochemistry*]. Moscow: Gosudarstvennoe izdatel'stvo geologicheskoi literaturу.

Schindewolf, O. H. 1993 [first published in German in 1950]. *Basic questions in paleontology : geologic time, organic evolution, and biological systematics*. Chicago: University of Chicago Press.

Schlesinger, W. H. 1991. *Biogeochemistry: An Analysis of Global Change*. San Diego: Academic Press.

Schopf, J. W. 1978. "The evolution of the earliest cells." *Scientific American* v. 239, pp. 110-138.

Schuchert, C. 1924. *Historical Geology*. [It was republished numerous times, coauthored with O. Dunbar]. New York: Wiley and Sons.

Schulz, E. 1937. "Das Farbstreifen-Sandwatt und seine Fauna, eine ökologische-biozönotische Untersuchung an der Nordsee" ["The farbstreifensandwatt and its fauna, an ecological-biozoonotic investigation in the North Sea"]. *Meereskundliche Arbeiten der Universität Kiel* v. 3, pp. 359-378.

Schwartzman, D., and T. Volk. 1989. "Biotic enhancement of weathering and the habitability of earth." *Nature* v. 340, pp. 457-460.

Seilacher, A. 1985. "Bivalve morphology and function." In *Mollusks: Notes for a Short Course*, Bottjer, D. J., C. S. Hickman, and P. D. Ward, eds. pp. 88-101.

Knoxville, Tennessee: University of Tennessee Department of Geological Sciences Studies in Geology 13.

Semper, K. 1881. *Animal Life as Affected by the Natural Conditions of Existence*. New York: D. Appleton and Company.

Shakhovskaya, A. D. 1988. "Khronika bol'shoi zhizni" ["Chronicle of a great life"]. In *Prometei 15, Vladimir Ivanovich Vernadskii, materialy k biografii, Istoriko-biograficheskii al'manakh serii "Zhisn' samechatel'nykh lyudei," 125-letiyu so dnya rozhdeniya V. I. Vernadskovo posvyashchaetsya* [*Prometheus 15, Vladimir Ivanovich Vernadsky, Biographical Materials. Historical-Biographical Almanac, Series "Lives of Famous People," In Recognition of the 125 Birthday of V. I. Vernadsky*], Mochalov, I. I., ed., pp. 33-85. Moscow: Molodaya gvardiya.

Shapley, H. 1927. *The Stars*. Chicago: American Library Association.

Shokalsky, Yu. M. 1917. *Okeanografia* [*Oceanography*]. Petrograd: Artistich. Saved. Tovarishch Marx.

Skoko, D., and J. Mokrovic. 1982. *Andrija Mohorovičič*. Zagreb: Skolska knjiga.

Smil, V. 1997. *Cycles of Life: Civilization and the Biosphere*. New York: Scientific American Library.

Smuts, J. C. 1926. *Holism and Evolution*. New York: Macmillan Company.

Snigirevskaya, N. S. 1988. "Pozdny devonvremya poyavlenya lesov kaka prirodnogo yavlenia" ["The first appearance of forests as a natural phenomenon in the Late Devonian"]. In *Stanovlenie i evolutsia kontinental'nykh biot* [*Formation and Evolution of Continental Biotas*], pp. 115-124. Leningrad: Nauka.

Soboleff, D. N. 1926. *Zemlia i Zhizn'* [*The Earth and Life*], Volume 1. Kiev.

Stapff, F. M. 1891. *Les eaux du tunnel du St. Gothard* [*The Waters of St. Gothard's Tunnel*]. Altenbourg: Edité par l'auteur.

Steffens, H. 1801. *Beytrage zur innern naturgeschichte der erde*. Freyberg: In Verlag der Crazichen Buchhandlung.

Stewart, W. D. G., G. P. Fitzgerald, and R. H. Burris. 1970. "Acetylene reduction assay for determination of phosphorus availability in Wisconsin Lakes." *Proceedings of the National Academy of Sciences (USA)* v. 66, pp. 1104-1111.

Stolz, J. F. 1983. "Fine structure of the stratified microbial community at Laguna Figueroa, Baja California, Mexico. I. Methods of the *in situ* study of the laminated sediments." *Precambrian Research* v. 20, pp. 479-492.

Stolz, J. F., and L. Margulis. 1984. "The stratified microbial community at Laguna Figueroa, Baja California, Mexico: A possible model for prePhanerozoic laminated microbial communities preserved in cherts." *Origins of Life* v. 14, pp. 671-679.

Stolz, J. F., L. Margulis, and R. Guardans. 1987. "La comunidad microbiana estratificada de la Laguna Figueroa, Baja California, México: un posible modelo de comunidades laminadas y microf—siles prefanerozoicos preservados en pedernales." *Studia Geologica Salmanticensia* v. 24, pp. 7-24.

Strakhov, N. M. 1978. "The conformity principle suggested by L. A. Zenkevich and its importance for the theory of ocean sedimentation." *Lithology and Mineral Resources* n. 4, pp. 124-133.

Suess, E. 1875. *Die Enstchung der Alpen* [*The Origin of the Alps*]. Vienna: W. Braunmüller.

———. 1904-1909. *The Face of the Earth* [Translated by Hertha B. C. Sollas]. Vol. 1, 1904; Vol. 2, 1905; Vol. 3, 1906; Vol. 4, 1909. Oxford: Clarendon Press [First published in German as *Das Antilitz der Erde*. Vol. 1, 1883; Vol. 2, 1888; Vol. 3, 1901; Vol. 4, 1904. Prague and Vienna: F. Temsky, and Vienna: Leipzig.].

Sugimura, Y., and Y. Suzuki. 1988. "A high temperature catalytic oxidation method for determination of non-volatile dissolved organic carbon in sea water by direct injection of liquid sample." *Marine Chemistry* v. 24, pp. 105-131.

Susiluoto, I. 1982. *The Origins and Development of Systems Thinking in the Soviet Union*. Helsinki, Suomalinen Tiedeakatemia: Annales Academiae Scientiarum Fenicae.

Tagliagambe, S. 1983. "The originality and importance of Vernadsky ideas." *Scientia* v. 118, pp. 523-535.

Tamman, G. G., and V. G. Khlopin. 1903. "Vliyanie vysokikh davleny na mikroorganizmy" ["The action of high pressure on microorganisms"]. *Vestnik obshchei gigieny sudebnoi i prakticheskoi mediciny* [*Herald of General, Forensic and Practical Medicine*] August-September, p. 51.

Tauson, L. V. 1977. *Geokhimicheskie tipy i potentsial'naya rudonosnost' granitoidov* [*Geochemical modes and the ore-bearing potential of granitoids*]. Moscow: Izdatel'stvo Nauka.

Termier, P. 1915. "Eduard Suess." *Revue générale des sciences pures et appliquées* v. 25, pp. 546-552 [English translation; Sketch of the life of Eduard Suess (1831-1914)." *Annual Report of the Smithsonian Institution for 1914*, pp. 709-718. Washington, D. C.].

———. 1922. *A la gloire de la Terre*. Paris: Desclée De Brouwer.

———. 1928. *La joie de connaître*. Paris: Desclée De Brouwer.

———. 1929. *La vocation desavant*. Paris: Desclée De Brouwer.

Termier, H., and G. Termier. 1952. *Histoire géologique de la biosphère*. Paris: Masson.

Thompson, D. W. 1952 [first published 1917]. *On Growth and Form, Second Edition*. Cambridge: Cambridge University Press.

Tikhomirov, V. V. 1969. In *Towards a History of Geology*, Schneer, C. J., ed., p. 357. Cambridge, Massachusetts: MIT Press.

Timiryazev, K. A. 1903. "The cosmic function of the green plant." *Proceedings of the Royal Society of London* v. 72, pp. 424-461.

Timonin, A. C. 1993. "Papytki sdelat' taksonomiyou bol'she biologicheskai (zamechanie Pavlinova 'Yest' li biologicheskieraznovidnosti')" ["Attempting to make taxonomy more biological (a comment Pavlinov's 'Is there the biological species')"]. *Zhurnal Obshchei Biologii* v. 54, pp. 369-372.

Tirring, F. 1925 [Thirring, H.]. "tensoranalyt. darst. d. elastiz. theor" ["On the tensor analytical representation of the theory of elasticity"]. *Physikalische Zeiteitschrift* v. 26, 5 s., pp. 518-522. September, 7.

Tissandier, G. 1887-1890. *Histoire des ballons et des aeronautes celebres*. Paris: H. Launette & Cie.

———. 1887. *Bibliographie aeronautique: catalogue de livres d'histoire, de science, de voyages et de fantaisie, traitant de la navigation aerienne ou des aerostats*. Paris: H. Launette & Cie.

Tomkeieff, S. I. 1944. "Geochemistry in the U.S.S.R." *Nature* v. 154, pp. 814-816.

———. 1945. "Prof. V. I. Vernadsky." *Nature* v. 155, p. 296.

Tort, P., ed. 1996. *Dictionnaire du Darwinism et de l'evolution: "Vernadskij"* v. 3, pp. 4439-4453. Paris: Presses Universitaires de France.

Toynbee, A. 1976. *Mankind and Mother Earth. A narrative history of the world*. Oxford: Oxford University Press.

Trees, C. C., R. R. Bidigare, and J. M. Brooks. 1986. "Distribution of chlorophylls and phaeopigments in the northwestern Atlantic Ocean." *Journal of Plankton Research* v. 8, pp. 447-458.

Trincher, K. S. 1965. *Biology and Information: Elements of Biological Thermodynamics* [Translated from Russian by E. S. Spiegelthal]. New York: Consultants Bureau.

Tropin, I. V., and E. Yu. Zolotukhina. 1994. "The time-course of heavy-metal accumulation by brown and red macroalgae." *Russian Journal of Plant Physiology* v. 41, pp. 267-273.

Trüper, H. G. 1982. "Microbial processes in the sulfur cycle through time." In *Mineral Deposits and the Evolution of the Biosphere*, Holland, H. D., and M. Schidlowski, eds., pp. 5-30. Berlin, Germany: Dahlem Workshop on Biospheric Evolution and Precambrian Metallogeny.

Use and Conservation of the Biosphere. 1970 [Proceedings of the Intergovernmental Conference of Experts on the Scientific Basis for Rational Use and Conservation of the Resources of the Biosphere, Paris. September 4-13, 1968]. Paris: Unesco.

Ushinsky, N. G. 1926. "K voprosu o proiskhozhdenii sernykh vod na poberesh'e Kaspiyskogo morya" ["On the problem of the origin of sulphuric waters on the coast of the Caspian Sea"]. *Azerbidzhan. Neftyanoe Kozyaistvo* [*Azerbaijan Oil Economics*] n. 8-9, pp. 83-84.

———. 1927. [Comment without title]. *Azerbidzhan. Neftyanoe Kozyaistvo* [*Azerbaijan Oil Economics*] n. 5, p. 69.

Vallery-Rodot, R. 1912. *The Life of Pasteur*. New York: Doubleday.

Van Hise, C. R. 1904. *A Treatise on Metamorphism*. Washington, D. C.: United States Geological Survey, Government Printing Office.

Varela, R. A., A. Cruzado, J. Tintore, and E. G. Ladona. 1992. "Modelling the deep-chlorophyll maximum: A coupled physical-biological approach." *Journal of Marine Research* v. 50, pp. 441-463.

Venrick, E. L., J. A. McGowan, and A. W. Mantayla. 1972. "Deep maxima of photosynthetic chlorophyll in the Pacific Ocean." *Fisheries Bulletin* v. 71, pp. 41-52.

Vernadsky, V. I. 1904. *Osnovy Kristallografii* [*The Fundamentals of Crystallography*]. Moscow: Publishing House of the Moscow Geological Institute.

———. 1923. "A plea for the establishment of a bio-geochemical laboratory." *The Marine Biological Station at Port Erin (Isle of Man) Annual Report, Transactions of the Liverpool Biological Society* v. 37, pp. 38-43.

———. 1924. *La Géochemie*. Paris: Félix Alcan.

———. 1926a. *Biosfera* [*The Biosphere*]. Leningrad: Nauka.

———. 1926b. *Izvestia Akad. Nauk SSSR Proceed.* p. 727; 1927, p. 241; *Revue Gen. des Sciences*, pp. 661, 700, 1926.

———. 1926c. *Nastavleniya dla opredeleniya geokhimicheskikh postoyannykh. 2. Opredeleniye geochimicheskoi inergii (velichiny D,V,e) nekotorykh grupp nasekomykh* [*Directions for the determination of geochemical constants. 2. Determination of geochemical energy (quantities D,V,e) for several groups of insects*]. Leningrad: Akademiya Nauk.

———. 1927. "Sur la dispersion des éléments chimiques." *Revue générale des sciences* v. 38, pp. 366-372.

———. 1929. *La Biosphère*. Paris: Félix Alcan.

———. 1930a [W. I. Wernadskij]. "Geochemie in ausgewahlten Kapiteln." *Akad. Verlagsgesellsch.* Leipzig, p. 370.

———. 1930b. "L'étude de la vie et la nouvelle physique." *Revue générale des Sciences* v. 41, pp. 695-712.

———. 1931. "Isotopes and living matter." *Chemical News* v. 142, pp. 35-36.

———. 1932. "Sur les conditions d'apparition de la vie sur terre." *Revue générale des Sciences* v. 43, pp. 503-514.

———. 1933a. [Title unknown]. *Z. Krist. Mineral. Petrog., Abt. B Mineral. Petrog. Mitt.* v. 44, p. 191.

———. 1933b [W. J. Vernadsky]. "Ozeanographie und Geochemie." *Zeitschrift fur Kristallographie, Mineralogie, und Petrographie Abteilugn B: Mineralogische und petrographische Mitteilungen* bd 44, heft 2/3, pp. 168-192.

———. 1934. "Le probléme du temps dans la science contemporaine." *Revue générale des Sciences* v. 45, pp. 550-558.

————. 1935. *Problemy biogeokhimii* [*Problems of Biogeochemistry*, 1st ed.]. Moscow and Leningrad: Izdatel'stvo AN SSSR.

————. 1937. "On the boundaries of the biosphere." *Izvestia Akademiya Nauk SSSR, ser. Geologich.* [*Proceedings of the Academy of Sciences of the USSR, ser. geol.*] n. 1, pp. 3-24.

————. 1938. "Biogeochemical role of the aluminum and silicon in soils." *Comptes Rendus (Doklady) de l'Academie des Sciences de l'USSR,* v. 21 (3), pp. 126-128.

————. 1939a. "On some fundamental problems of biogeochemistry." *Travaux du Laboratoire Bioge—chemique de l'Académie des Sciences de l'URSS* v. 5, pp. 5-17.

————. 1939b. "On some fundamental problems of biogeochemistry." *Travaux du Laboratoire Biogéochimique de l'Académie des Sciences de l'URSS* v. 21, pp. 5-17.

————. 1940. *Biogeokhimicheskie Ocherki* [*Biogeochemical Essays*]. Moscow and Leningrad: Izdatel'stvo Akademiya Nauk SSSR.

————. 1942. "On geological envelopes of the Earth as a planet." *Izvestia Akademiya Nauk SSSR, ser. geogr. geofiz.* n. 6, pp. 251-262.

————. 1944. "Problems of biochemistry, II." *Transactions of the Connecticut Academy of Arts & Sciences* v. 36, pp. 483-517.

————. 1945. "The biosphere and the noösphere." *American Scientist* v. 33, pp. 1-12.

————. 1954. "Izbrannye sochineniya" ["Selected Works"]. *Izdatel'stvo Akad. Nauk SSSR, Moscow* v. 1.

————. 1959. *Istoriya mineralov zemnoi kory, 1., Izbr. soch., IV.* [*History of the mineralogy of the Earth's crust, v. 1*]. Moscow: Izdatel'stvo Akademiya Nauk SSSR.

————. 1960. *Biosfera*. Beograd: Kultura.

————. 1965. *Khimicheskoe stroenie biosfery Zemli i ee okruzhenie* [*The Chemical Structure of the Earth's Biosphere and its Surroundings*]. Moscow: Nauka.

————. 1967. *Biosfera* [*The Biosphere*, footnotes above ascribed to A. I. Perelman are translated from this edition by D. Langmuir]. Moscow: Mysl'.

————. 1977. *Scientific Thought as a Planetary Phenomenon* [Translated by B. A. Starostin]. Moscow: Nongovernmental Ecological V. I. Vernadsky Foundation.

————. 1978. *Zhivoe Veshchestvo* [*Living Matter*]. Moscow: Nauka.

————. 1980. "Problemy Biolgeokhimii" ["Problems of Biogeochemistry"]. *Trudy Biogeokhim. lab.* [*Transactions of the Biogeochemistry Laboratory*] v. 16, pp. 1-320.

————. 1985. *Pis'ma V.I. Vernadskogo A.E. Fersmanu* [*The Vernadsky-Fersman Correspondence*]. Compiled by N. V. Filioppova. Moscow: Nauka.

————. 1986. *The Biosphere* [abridged version]. Oracle, Arizona: Synergetic Press.

————. 1988 [Vernadskii]. "Pamyati M. V. Lomonosova" ["Memorial to M. V. Lomonosov"]. In *Prometei 15, Vladimir Ivanovich Vernadskii, materialy k biografii, Istoriko-biograficheskii al'manakh serii "Zhisn' samechatel'nykh lyudei," 125-letiyu so dnya rozhdeniya V. I. Vernadskovo posvyashchaetsya* [*Prometheus 15, Vladimir Ivanovich Vernadsky, Biographical Materials. Historical-Biographical Almanac, Series "Lives of Famous People," In Recognition of the 125 Birthday of V. I. Vernadsky*], Mochalov, I. I., ed., pp.326-328. Moscow: Molodaya gvardiya.

————. 1989. *Biosfera i Noosfera* [*Biosphere and Noösphere*]. Moscow: Nauka.

————. 1991. *Nauchnaya Mysl'kak Planetnoe Yavlenie* [*Scientific Thought as a Planetary Phenomenon*]. Mosco: Nauka.

————. 1992. *Trudy po Biogeokhimii i Geokhimii Pochv* [*Contributions Con-*

cerning the Biogeochemistry and Geochemistry of Soils]. Moscow: Nauka.

———. 1993. La Biosfera. Como, Italy: Red Edisioni.

———. 1994. Zhivoe Veshchestvo i Biosfera [Living Matter and the Biosphere]. Moskva: Nauka.

———. 1995 [Vernadskij]. Pensieri filosofici di un naturalista. Rome: Teknos Edizioni.

Vinogradov, A. P. 1963. "Development of V. I. Vernadsky's Ideas." Soviet Soil Science v. 8, pp. 727-732.

Vinogradsky, S. 1887. "Uber Schwefwlbacterien." Bot. Ztg. 45, pp. 489-507, 513-523, 529-539, 545-559, 569-576, 585-594, and 606-610.

———. 1888a [Winograsky]. Beiträge zur morphologie und physiologie der bacterien [Contributions to the morphology and physiology of bacteria]. Liepzig: Zwickau.

———. 1888b. "Ueber eisenbacterien." Bot. Ztg. 46, pp. 261-270.

———. 1895 [Winogradsky]. "Recherches sur l'assimilation de' l'azote libre de l'atmosphere par les microbes." Arch. des sci. biolog. St. Petersburg v. iii, pp. 297-352.

———. 1989. "Recherches physiologiques sur les sulfobacteries." Ann. Inst. Pasteur 3, pp. 49-60.

Voitkevich, G. V., A. E. Miroshnikov, A. S. Povarennykh, and V. G. Prokhorov. 1970. Kratkii spravochnik po geokhimii [Brief handbook of geochemistry]. Moscow: Izdatel'stvo "Nedra".

Weber, B. H. Et. al., eds. 1988. Entropy, Information, and Evolution: New Perspectives on Physical and Biological Evolution. Cambridge, MA: MIT Press.

Weinberg, S. 1988. The First Three Minutes: A Modern View of the Origin of the Universe. New York: Basic Books.

Wentworth, W. E., and S. Jules Ladner. 1972. Fundamentals of Physical Chemistry. Belmont, California: Wadsworth Publishing Company.

Westbroek, P. 1991. Life as a Geological Force. New York: W. W. Norton & Company.

Weyl, P. K. 1966. "Environmental satibility of the earth's surface—chemical consideration." Geochimica et Cosmochimica Acta v. 30, pp. 663-679.

Whitehead, A. N. 1926. Science and the Modern World. New York: Macmillan Company.

———. 1978. Process and Reality: An Essay in Cosmology [Gifford Lectures, 1927-1928]. New York: Free Press.

Wicken, J. S. 1987. Evolution, Thermodynamics, and Information: Extending the Darwinian Program. New York: Oxford University Press.

Widdel, F., S. Schnell, S. Heising, A. Ehrenreich, B. Assmus, and B. Schink. 1993. "Ferrous iron oxidation by anoxygenic phototrophic bacteria." Nature v. 362, pp. 834-836.

Wiener, N. 1948. Cybernetics or Control and Communication in the Animal and the Machine. New York: John Wiley & Sons, Inc.

Wiesner, I. 1877. Die Entstehung des Chlorophyl der Pflanze. Eine physiologische Untersuchung. Vienna: Hölder.

Williams, G. R. 1997. The Molecular Biology of Gaia. New York: Columbia University Press.

Williams, H., F. J. Turner, and C. M. Gilbert. 1982. Petrography: An Introduction to the Study of Rocks in Thin Sections, Second Edition. New York: W. H. Freeman and Company.

Williamson, E. D., and L. H. Adams. 1925. "Density distribution in the Earth." Journal of the Washington Academy of Sciences v. 13, pp. 413-428.

Wood, J. A.. 1968. Meteorites and the Origin of Planets. New York: McGraw-Hill.

Yanshin, A. L., and F. T. Yanshina. 1988. "The scientific heritage of Vladimir Vernadsky." Impact of Science on Society v. 151, pp. 283-296.

————. ed. 1989. "Znachenie nauchnogo naslediya V. I. Vernadskogo dla sovremennosti." *Nauchnoe i sotsial'noe znachenie deyatel'nosti V. I. Vernadskogo [Scientific and Social Significance of V. I. Vernadsky's Activity].* Leningrad: Nauka Leningradskoe Otdelenie.

Young, J. Z. 1971. *An Introduction to the Study of Man.* Oxford, Great Britain: Clarendon Press.

Zaninetti, L. 1978. "Les 'fossiles vivants'; Note a l'occasion du 40eme anniversaire de la decouverte du premier 'Coleacanthe', le 22 decembre 1938" ["Living fossils; note on the 40th anniversary of the discovery of the first 'coelacanth', December 22, 1938"]. *Notes du Laboratoire de Paleontologie* v. 3, pp. 19-21.

Zettnow, E. 1912. "Ueber ein vorkommen von sehr widerstandsfahigen bacillensporen." *Centralbl. Bakt. Jena* abt. 1, 66, originale, pp. 131-137.

Zimmer, C. 1996. "The light at the bottom of the sea." *Discover* v. 17, pp. 62-73.

Acknowledgments

It's unusual for a book's publisher to write its acknowledgments. This edition of *The Biosphere*, however, drew on the talents of such disparate participants, separated by many miles and decades, that no one else is in a position to thank (or even know of) everyone to whom thanks is due. So, by default, this pleasant task falls to me.

This book's publication was made possible, first and foremost, by the advocacy, counsel, and genius of Lynn Margulis. She has inspired a generation of students, challenged her colleagues to see deeper into the nature of living matter, and been a model of excellence and daring to all those who are privileged to know her. She is a worthy successor to Vernadsky.

Jacob Needleman, John Pentland, and the other members of the Far West Institute had the prescience in the 1970s to commission a translation of *The Biosphere*. It is that translation, somewhat revised, which has become this book. David Langmuir, the original translator, generously made available all his notes, including summaries of his conversations with Evelyn Hutchinson regarding Vernadsky and his ideas.

I am grateful to all those mentioned in the Foreword for their wisdom and aid in obtaining written information about Vernadsky's biosphere. Andrei Lapo, the great Vernadsky scholar, answered many questions about Vernadsky's life and his concept of the biosphere. Wolfgang Krumbien and his colleague, George Levit, suggested valuable corrections to the original translation. Jacques Grinevald, whose Introduction graces this book, contributed insights into Vernadksy's place in the history of science. I. A. Perelman made available annotations he had prepared for a 1967 Russian-language edition of *The Biosphere*, some of which have been translated and reprinted in this edition. Others whose comments, insights, and work were of value for the completion of this project include Connie Barlow, Stewart Brand, Michael Chapman, Al Coons, Ludovico Galleni, Loren Graham, David Grinspoon, Guenzel Guegamian, James Lovelock, Dianna McMenamin, Louis Orcel, Sergei

Ostroumov, Cheryl Peach, Nicholas Polunin, Donna Reppard, Stephen Rowland, Dorion Sagan, Paul Samson, Richard Sandor, Eric Schneider, David Schwartzman, George Theokritov, Francisco Varela, Tyler Volk, Tom Wakefield and Sandy Ward. A special thanks is due Joe Scamardella who in a Herculean effort tracked down many of the obscure 19th and early 20th century citations. Sharon Dornhoff is acknowledged for her faithful transcription of the original translation and Gary Halsey for his preparation of the index.

José Conde, graphic artist extraordinaire, has given this edition of The Biosphere its elegant jacket and interior design.

At Springer-Verlag, William Frucht, who believed in The Biosphere from the first day, contributed mightily to each and every stage of its editing and production; Jerry Lyons, an editor who has changed the face of science publishing in this country, provided much crucial support; and Teresa Shields lent her editorial, Natalie Johnson her production, and Karen Phillips her design expertise; and at Nevraumont Publishing, I thank Simone Nevraumont for her aid in preparation of the manuscript and Ann J. Perrini for her sage advice.

Last but not least, Mark McMenamin deserves thanks for his rigorous review and revision of the original translation and for his illuminating notes on the antecedents, intent, and successors to Vernadsky's world view.

Vladimir Vernadsky, a modest man, might well be surprised by all the attention accorded to his book and himself. I hope he would be pleased, after a 70-year gestation, with the appearance of this English-language edition of the complete text of his great work.

Peter N. Nevraumont
October 1997

Index

я пытался выяснить. что ч

крайней мере в 10 раз,

е 0,25% солнечной энерги

запасе — в живом веще

обом термодинамическом

ля косной материи биосф

ичества вещества, посто

низмы, большие количес

ислорода (около $1,5 \times 10^{21}$

ражается в меньших чис.

авливающиеся размножен

как указывалось (§ 45),

в, во много раз превышаю

многократные числа

с его учесть, энергетическ

ы не можем их понять, если

ке планетных явлений, не обрат

терии, к ее атомам, к их измен

й области быстро накапливаю

заченные теоретической мыслью

наваться. Они не всегда могут б

заны, и выводы из них обычно не

Но огромное значение этих явл

вые факты должны теперь же уч

виях. Три области явлений мо

особое положение элементов з

ме. 2) их сложность и 3) неравно

Так, в массе земной коры рез

нты, отвечающие четным ато

ъяснить это явление геологичес

 не можем. К тому же нем

мое явление выражено еще более

мле космических тел, доступн

учению — для метеоритов (Гарки

Область других фактов являе